SAGE
50
YEARS

Gender Issues in Water and Sanitation Programmes

Thank you for choosing a SAGE product! If you have any comment, observation or feedback, I would like to personally hear from you. Please write to me at contactceo@sagepub.in

—Vivek Mehra, Managing Director and CEO,
SAGE Publications India Pvt Ltd, New Delhi

Bulk Sales

SAGE India offers special discounts for purchase of books in bulk. We also make available special imprints and excerpts from our books on demand.

For orders and enquiries, write to us at

Marketing Department
SAGE Publications India Pvt Ltd
B1/I-1, Mohan Cooperative Industrial Area
Mathura Road, Post Bag 7
New Delhi 110044, India
E-mail us at marketing@sagepub.in

Get to know more about SAGE, be invited to SAGE events, get on our mailing list. Write today to marketing@sagepub.in

This book is also available as an e-book.

Gender Issues in Water and Sanitation Programmes

Lessons from India

Edited by
Aidan A. Cronin
Pradeep K. Mehta
Anjal Prakash

First published in 2015 by

SAGE Publications India Pvt Ltd
B1/I-1 Mohan Cooperative Industrial Area
Mathura Road, New Delhi 110 044, India
www.sagepub.in

SAGE Publications Inc
2455 Teller Road
Thousand Oaks, California 91320, USA

SAGE Publications Ltd
1 Oliver's Yard, 55 City Road
London EC1Y 1SP, United Kingdom

SAGE Publications Asia-Pacific Pte Ltd
3 Church Street
#10-04 Samsung Hub
Singapore 049483

Published by Vivek Mehra for SAGE Publications India Pvt. Ltd, typeset in 10/13pt Berkeley by Diligent Typesetter, Delhi and printed at Sai Print-o-Pack, New Delhi.

Library of Congress Cataloging-in-Publication Data Available

ISBN: 978-93-515-0065-0 (HB)

The SAGE Team: Rudra Narayan, Alekha Chandra Jena, Nand Kumar Jha and Vinitha Nair

Contents

SECTION 2
Case Studies: Water

SECTION 3
Case Studies: Sanitation

List of Tables

List of Figures

List of Boxes

Foreword

For many years, attempts to reduce poverty have included water for various purposes, whilst sanitation has received due attention only since the last 10 to 15 years. Nevertheless, even now, in India and in many other parts of the world, many women and girls walk long distances to fetch water carrying pots and buckets and to find suitable places for relieving themselves, daily. Not all development efforts are adequate, and it appears to be difficult to keep up with the increasing population. Case studies that describe the context, the effort made, the successes and the problems are important for others to learn from. The book you have just now opened covers many different examples from India, which are also interesting for those who try to improve water, sanitation and hygiene (WASH) facilities for poor women and men worldwide.

In 1963, almost 50 years ago, I first came to live in South Asia, and I learned to use water from irrigation canals, and to do without toilet, hiding in the dark of the night, with all its invisible dangers. In 2013, the Gender Team of the Sujal Project[1] did research in nine villages, asking women about their work related to water and sanitation. These villages hardly had a few toilets each, and not one anybody wanted to get close to. Since 1963, lots of efforts have been made towards water development, but so many people, women and men, elderly and disabled people, still have to find their way in the dark of the night. This contradiction must be better understood.

In order to contribute to the discussion on practices and initiatives that have proven to be effective in promoting sustainable access to water, sanitation and hygiene, by integrating the interests of the people who are different by gender, age, ethnicity, ability, caste and socio-economic position, a National Conference on Women-led Water Management was organised by the S M Sehgal Foundation in partnership with UNICEF India in November 2012. The conference focused on strategies towards sustainable WASH in rural India and brought together a diverse and knowledgeable cadre of governmental organisations (state and national spanning several departments), non-governmental organisations, specialised agencies, businesses and education institutions. The conference shared success stories related to women's leadership and participation in water management and sanitation, needs assessment, planning, decision-making, implementation, monitoring and social audit. Innovative approaches to raise women's dignity and eliminate water-related drudgery work of women and girls were presented. The focus on equity and inclusion, capacity building, policy and good governance was strong.

The opening discussions looked at the larger gender concerns such as the different perception one generally has of men and women farmers, where women are rarely seen as productive farmers, at best as helping their husbands, and therefore often alienated from rights to land and water. This, in times where many men leave their plots of land to their wives, seeking employment in cities, results in unnecessary low food production. Further, it was conveyed that the centralised programmes in India, including WASH schemes, often result in disempowering the villagers they are meant to support by diminishing participation and by taking away the locus of control from the water users themselves to a technological fix. Participants of the National Conference stressed the need to recognise that gender does not only mean the differences between women and men, but also that women are differentiated by class, caste, religion, age and ethnicity, as men are, and therefore a more nuanced understanding of women in WASH issues is necessary. There was a call for the need for institutional reform to acknowledge women's opinions, actions and steering in water and development, making voices of women of different background heard.

Various field experiences were shared, and the key question was raised as to what extent gender barriers have been overcome, beyond the project-related objectives, targets and time frame. Most projects require the participation of women, and some will be successful in reaching their aims. Yet, that alone does not necessarily ensure that gender barriers are broken and relations between women and men have become more equal; on the contrary, it often means that women's workload has increased. Hence, there is a need to apply a larger framework of socio-cultural, economic, political and physical empowerment of women (and men) which remains a key challenge in gender inclusion. Some of the indicators that describe the extent of empowerment are quantifiable and measurable, while some others are not. It is important to document both quantitative and qualitative information for gender-disaggregated data monitoring. Data are needed for policy makers to influence policies and the budget at different levels. Also, working for development, we need to be able to locate and identify which larger gender barriers have been addressed either by design or by default and what more can be done.

A critical question is which interventions and approaches actually deliver on women-led water management, especially in large and well-funded government programmes. The government agencies consisting of engineers of the State Public Health and Engineering (PHED) or Rural Development Departments do not have sufficient numbers of field staff who are skilled in undertaking water and sanitation programmes with a gender empowerment focus. Often, Information Education Communication materials such as posters, commercials and short films are used, but without inter-personal communication, these have little impact on sustainable behavioural change or changes in perceptions around gender. We must build on participatory approaches which work with capacity development tailored at different levels of stakeholders. Where a solid process, with women playing a central role, was followed then the water committees in the villages could manage the project once it was completed by the state government.

There is still a long way to go for all women and men to have access to appropriate facilities, and for women to become equal initiators and decision makers in the WASH sector in India. But this book shows that change is possible. It collates good experiences from right across the country and is an excellent contribution to filling the large gap in the systematic documentation of work on gender relations in the water and sanitation sector. The book will help accelerate progress towards this goal by providing pointers and promising practices for government, civil society, academia and above all the women water users to follow.

Joke Muylwijk
Executive Director, Gender and Water Alliance

Note

1. Sujal, EU-funded programme, led by VRUTTI called: 'Establish and demonstrate a people and Panchayat led equitable water governance model for Sustainable Economic Development in three Agro-Ecological Zones of India', in which the Gender and Water Alliance contributes with capacity building. The Gender team consists of Joke Muylwijk, Jhansi Rani Ghanta and Meena Bilgi.

1

Introduction: Achieving the Desired Gender Outcome in Water and Sanitation

Anjal Prakash, Aidan A. Cronin and Pradeep K. Mehta

> *Ending gender based inequities, discrimination and violence faced by girls and women must be accorded the highest priority ... certain essential interventions ... need to be made ... such as provision of sanitation facilities, including construction of toilets with water facility in schools. The effort to promote women's health cannot be without participation of men; hence, imaginative programs to draw men into taking part in their health seeking behaviour and practices must be devised.*
>
> (Planning Commission, 2011: 88–89)

Gender equality is central and critical to economic and human development. Yet there is no country in the world where gender inequality does not exist. In developing countries, gender disparity compounds with poverty, inequity, poor health and access to critical and basic services. A recent report—*The World's Women 2010: Trends and Statistics* (United Nations, 2010)—presents global statistics and analysis on the status of women and men, highlighting the current situation and changes over time. The gender-related disparities that are stated are appalling. Women account for two-thirds of the world's 774 million adult illiterates in 2010, a proportion that is unchanged over the past two decades. Further, the report shows that gender disparities in adult literacy rates remain wide in most regions of the world. While the overall progress in primary education in the past decade is encouraging, major barriers stand in the way of progress: 72 million children—54 per cent of them girls—are out of school. Globally, women's participation in the labour market has remained steady over the period from 1990 to 2010, hovering around 52 per cent. However, women continue to bear most of the responsibilities for the home, caring for children and other dependent household members, preparing meals and doing other housework. Across continents, women spend at least twice as much time as men on unpaid domestic work.

For this book, the term gender refers to

> culturally based expectations of the roles and behaviours of men and women. The term distinguishes the socially constructed from the biologically determined aspects of being male and female. Sex identifies the biological difference between men and women, whereas gender identifies the social relations between men and women. It therefore refers not to men and women but to the relationship between them, and the way this is socially constructed. Gender relations are contextually specific and often change in response to altering circumstances. (Moser, 1993: 230)

The gender disparity in socio-economic and political areas gets reflected in developmental programmes leading to skewed gender outcomes. This disparity in socio-political arenas also gets reflected in the water, sanitation and hygiene (WASH) sector. In fact, gender issues in water and sanitation are a subset of larger gender concerns in the society. Water is clearly a male-dominated sector and a subject articulated in masculine terms and infrastructure reinforcing the rules of patriarchy and prevailing social relations (Cleaver, 1998; Meinzen-Dick and Zwarteveen, 1998; Zwarteveen, 2008, 2011). The lopsided understanding of gender and its interrelation with water and sanitation issues have led to gender disparity in access to water supply and sanitation (Prakash et al., 2012). In most societies, women have primary responsibility for the management of household water supply, sanitation and health. Water is necessary not only for drinking, but also for food production and preparation, for taking care of domestic animals and the sick, personal hygiene, cleaning, washing and waste disposal. Because of their dependence on water resources, women have accumulated considerable knowledge about water resources, including location, quality and storage methods. However, efforts geared towards improving the management of the world's finite water resources and extending access to safe drinking water and adequate sanitation often overlook the central role of women in water management (UN Water, 2006).

This book fills existing gaps in knowledge. It illustrates how to get desired gender outcomes in WASH programmes by providing real-life case studies from different regions of India. This book is necessary and timely for two reasons. Firstly, the published literature on WASH challenges in India is distorted. Apart from few recent case studies focussing on different parts of India (Baby and Reddy, 2013; Cullet, 2009; Kumar et al.,

2011; Pattanayak et al., 2010; Prakash and Sama, 2008; Sangameswaran, 2010; Starkl et al., 2013), there is a lack of the comprehensive all-India documentation of WASH issues. Secondly, the literature covering gender issues in WASH is even more skewed. Apart from some of the seminal works that focus on larger gender and water issues (Coles and Wallace, 2005; Lahiri-Dutt, 2006; Lahiri-Dutt and Wasson, 2008; Zwarteveen et al., 2013), the recent gendered works of literature emphasising WASH issues in India are very few (Ahmed, 2005; Joshi, 2005; O'Reilly, 2006; Rao, 2008). This book attempts to fill this important gap to understand how we can get desired gender outcomes in WASH programmes with a special emphasis on India. The focus of the book is primarily on water and sanitation concerns, whereas gender and hygiene–related issues remain a gap here that requires further analysis. The editors feel that hygiene issues are absolutely intrinsic to water and sanitation outcomes but due to lack of focused work on hygiene issues in India, this book is unable to make that link as strongly as is needed and is an area that more work is needed for the future.

This introductory chapter provides the outline of why skewed gender and equity outcomes occur in WASH programmes. Providing the current status and understanding the Indian context, this chapter reviews the problems of gender imbalance by interconnecting gender and WASH linkages and is divided into three sections. Section 1 focuses on 'Gender and WASH Problématique' that forms the background for the cases in this book. Section 2 provides the 'Context of Gender and WASH in India' through basic facts and outlines key statistics. Section 3 gives an overview of the chapters in this book.

1.1 The Gender and WASH Problématique

Rural and urban women of all age groups are engaged in collection of water for household needs, including water for livestock. This wields a physical and health burden on women, particularly pregnant women (UNICEF, 2011). In addition, poor water quality and inadequate sanitation negatively impact their food security, levels of nutrition and livelihood choices. It also significantly lessens the time available to women and girls to attend to other development issues. Girl children's educational and overall self-development status suffers a serious setback in a society where they are considered inferior to boy children as they are involved in water collection and other household chores constrained by water supply. The brunt of inadequate sanitation facilities is also borne by the women; in both rural and urban areas, travelling long distances in search of privacy to relieve themselves exposes women to both lack of security and health hazards. Police in Bihar estimate that over 40 per cent of rapes in the state occur as women go for open defecation (*Times of India*, 2013).

Furthermore, the introduction of technology has not led to a significant change in gender stereotypes. For example, while the general impression is that the introduction of hand pumps have reduced the burden of women in terms of physical labour of fetching water from long distances, in some cases the intended gender outcomes have not been realised. For instance, Narain (2003) reports that the availability of 24-hour water supply by hand pumps was accompanied by change in aspirations of the village men who then started bathing in their own houses, an activity that was earlier done at the village pond, thus increasing the burden on women to fetch extra water for this purpose. A review of the national sanitation programmes (Total Sanitation Campaign prior to 2012) and rural water supply (Swajaldhara) schemes in 10 districts in the state of Uttar Pradesh has revealed that in some sample villages women were using toilets more than men, who still defecated in the open (Chand et al., 2004). Furthermore, women who are encouraged to promote hygienic, health-supporting behaviours are better able to improve the lives of children and can influence others in the community to adopt such behaviours. UNICEF (2011) also indicates that children who are influenced during the formative years are more likely to teach their own children healthy habits.

The extent of these issues is compounded when one examines the professionals involved with improving the gender situation in the WASH sector. They are primarily men (UNICEF, FAO and SaciWATERs, 2013) with gender stereotypes holding back progress (Zwarteveen, 2011). Clear inequities exist in WASH provision in India that require dedicated interventions for reaching marginalised groups. Gender cuts across all the issues as it intersects with class and caste and produces layered social hierarchies that impinge on one's access to and control over precious resources such as water. Equally important is gender in accelerating sanitation. With participatory communication and proper process, these interventions can impact WASH-borne diseases as well as lead to women empowerment and gendered outcomes (e.g. Lala et al., in this volume). However, for now gender and water issues still remain at the level of rhetoric for the want of a broad-based and shared understanding, without any support from the ground and especially when it comes to data on changing gender and social relations (UNICEF, FAO and SaciWATERs, 2013). Evidence is urgently required to inform the debate—gender-disaggregated data are a serious gap that needs to be taken up as priority for any progress to be made in mainstreaming gender in the WASH sector. The bureaucratic set-up that manages water also suffers from serious inadequacies with respect to gender mainstreaming in their institutions. Without addressing these issues, well-meaning gender inclusion efforts will not lead to logical and intended outcomes.

Water and sanitation programmes have followed a narrow focus on gender-sensitive mechanisms of water delivery. This has mainly happened through committees, tariffs and technologies, which by themselves cannot ensure gender-equitable outcomes. In order to achieve gender parity, there is a need for a broader understanding of the ways in which societal resources are allocated through economic policies and legislations and to shape mechanisms in particular ways. Furthermore, it is critical to consider how different people are able to influence the outcomes of particular governance arrangements to produce gendered outcomes (for health and well-being, access and livelihoods, and for political voice, ultimately resulting in increased gender empowerment).

1.2 Context of Gender and WASH in India

The WASH sector in India is dominated by two national programmes—the National Drinking Water Programme and the *Nirmal Bharat Abhiyan* (or clean India campaign)—that address sanitation. These are shaped by policy and funding outlays from the Government of India and are implemented by the state governments, which are also expected to contribute in terms of both financial and human resources. State implementation is built upon coordinating three layers of government—State (for monitoring and capacity building), District (for programme implementation) and Local Government at the sub-district level. A positive development at the national level was the culling out of National Ministry of Drinking Water and Sanitation (MDWS) to focus primarily on improving India's water and sanitation indicators. However, what did not happen was a similar culling of roles and responsibilities at the state and sub-state levels. This has resulted in water and sanitation being implemented by a wide and diverse range of state departments, from Rural Development to *Panchayat Raj* (Local Government) to Public Health and Engineering and other variants of these. This has not helped focus on cross-cutting issues such as addressing gender or social inequities in WASH. Neither has the lack of convergence with other ministries working on related issues such as Ministry of Health, Ministry of Education and especially in terms of gender with Ministry of Women and Child Development. Any convergence that has happened has been driven by champions and not institutionally led or sustained.

Disparities in WASH service provision exist across states and districts, rural-urban divides and religious and social grouping, including among politically recognised excluded classes such as scheduled castes (SCs) and scheduled tribes (STs). Disparities in access to WASH services are seen across different social classes with the poorest least served (UNICEF, 2012a). Sixty-five per cent of the richest quintile, that is, the richest 20 per cent of the population of India, have piped water on premises, while only 2 per cent of the poorest quintile[1] have piped water access; in rural areas, 32 per cent of the richest quintile have piped water on premises, while only 1 per cent of the poorest quintile have it (NFHS, 2006). A household survey available shows that 15 per cent of the poorest quintile of Indians have access to a toilet, while 58 per cent of the richest have access (NSSO, 2010).[2]

Caste- and tribe-based inequity has resulted in deprivation due to social barriers in access and effective use—only 24 per cent of tribal populations have access to piped water, whereas 52 per cent of 'others' have piped water access. The urban–rural divide is also evident; 65 per cent of rural India defecate in the open, while only 11 per cent of urban areas defecate in the open. Due to the environmental and societal needs, urban areas appear to reflect better WASH coverage. However, the urban poor are a discriminated group within the urban segment (UNICEF,

FAO and SaciWATERs, 2013). Geographic inequities exist with richer states and districts closer to state capitals in general being better served (UNICEF, 2012b; UNICEF and WHO, 2012). Monitoring of the more remote areas is weaker and so it is more difficult to determine the WASH service reach and gaps.

The young are also affected by WASH inequities. Poor WASH in school facilities leads to high dropout rates by young and adolescent girls, particularly during menstruation. Even when sanitation facilities are available, there are some challenges involved in ensuring all students have access to them; for instance, children from Dalit families are often discriminated against when it comes to using toilets (Jacob et al., 2009). Even more serious is the issue of stunting, contributing to over a third of under-five deaths globally (GLAAS, 2012; UNICEF, 2012c). India has the largest number of stunted children in the world with over 40 per cent moderately or severely underweight (NFHS, 2006). As stunting depends on food intake, general health status and physical environment, WASH is critically linked to all three (Fewtrell et al., 2007). Poor WASH is inextricably linked to high stunting in populations (Esrey, 1996).

In Alwar district in the state of Rajasthan, the school sanitation programme increased girls' enrolment by one-third, leading to a 25 per cent improvement in academic performance for both boys and girls (UNICEF, 2011). The needs of disabled people in developing countries are consistently overlooked when it comes to WASH, with severe consequences on the health, dignity, education and employment of disabled people and their caregivers (WaterAid, 2011). In India, poor-quality water both causes and compounds disability (Cronin, 2013), while grossly inadequate sanitation facilities for the disabled are evident throughout the country (WaterAid, 2013).

Gender is, perhaps, the largest inequity issue in the WASH sector in India as it not only cross-cuts all the above issues but also creates major inequities in its own right. This is not to oversimplify and say, for example, that women can be seen as a homogenous group. Women are also not equally divided by caste, poverty, geography and wealth (Joshi, 2011). However, to unlock the true potential for gender to strengthen the WASH sector in India, we must first understand the issue and then view the strategic interventions necessary to achieve this impact. We now unpack this further.

1.2.1 Gender and WASH in India—Progress to Date

One of the major changes in the water sector over the last few decades has been the enhanced thrust on institutional reforms, including the increased recognition of the bottom-up approach to management as against the techno-centric top-down ones. Gender is an important component of this policy; the recognition of the greater need of understanding women's differential needs is critical (e.g. MDWS, 2011). A need to integrate gender concerns in the water policy discourse stems from the fact that women are the primary collectors of water and also responsible for health, hygiene and sanitation (of which water is an integral part) at the household level.

The recent increase of rhetorical support around 'gender mainstreaming' by government and civil society is welcome but it also needs to be transformed into affirmative action. Though gender is getting increased visibility and its importance being recognised more and more, for example, in the Planning Commission's preparations for the 12th Five Year Plan, the policy must result in ground impact, especially among the most disadvantaged. Indeed, policy makers have used many approaches to address gender concerns (Prakash, 2012). For example, the 'efficiency' approach is based on the premise that if given an opportunity, women can be as efficient in delivering targets as men, or, in some cases, even better. There is also the 'needs' approach, advocating the fact that women have very specific needs related to water. Finally, there is the 'equity' approach, which makes a case for mainstreaming gender concerns in policies to address equity issues (Cap-Net, SOPPECOM and WWN, 2007).

However, despite policy intentions to mainstream gender issues in WASH, there is a lack of adequate and systematic data (e.g. Seager, 2010). Without the support of reliable gender-disaggregated data, it is difficult to track changes in gender relations or gender impacts across India. Though some progress is being seen with new household surveys, for example, NSSO 69th Round in 2012, much more can still be done.

As per the Government of India guidelines, the Village Water and Sanitation Committee members should be selected to represent various groups of society. Fifty per cent of these committees should be women, especially those belonging to SC, ST and OBC categories (MDWS, 2011). Women's representation in the water sector has come a long way from being gender-biased. Women are now viewed as a disadvantaged group with respect to access and decision-making processes around water. However, the effectiveness of this and its translation into increased empowerment is not clear (Bastola, Chapter 7 of this volume). Approaches have often failed to recognise the fact that women can contribute immensely to the decision-making process and therefore they should not only be seen as potential beneficiaries but also as key stakeholders and important decision-making actors. Many such positive examples are presented in this book.

1.2.2 Gender Inclusive Governance

It is perceived that India has some of the best policies but lacks proper implementation and this is where most of the major challenges lie. Proper implementation of programmes/schemes and collaborative functioning of the key institutions involved the three-tier local governance institutions, the line agencies and the non-government organisation (NGO)-initiated/ supported self-help groups (SHGs); the latter having no state-given functional, administrative or financial powers and no direct links with *Panchayati Raj* Institutions (PRIs) or line agencies but being the fulcrum for most of the successes in the case studies. Effective governance is central to the goals for reducing poverty and inequality and providing equal opportunities (Shah and Kulkarni, 2013). Good governance is a normative term that denotes how accountable, transparent, inclusive and responsive government institutions are to their citizens (Brody, 2009), and sustainable WASH delivery is a direct outcome of good governance process. There are two fundamental questions—first, how to make governance work and second how could it be more gender-sensitive? 'Making governance gender-sensitive requires more than adding women in Parliaments, but this is one place to start. Gender-sensitive reforms in national and local government—in the form of electoral quota systems and the establishment of women's ministries—have helped to achieve a better gender balance' (Brody, 2009: 2). Answering the second question borders on the issue of gender responsive budgeting (GRB), which has a scope for engendering governance but is yet to pick up in India. Mishra and Sinha (2012) reveal that the two prime strategies adopted by the Government of India for institutionalising GRB are Gender Budget Statement and Gender Budgeting Cells to highlight what has gone wrong, and what needs to be fixed. The authors argue that low priority is given to gender agenda and the gender cell lack coordinating mechanism to translate into better outcomes for the women of the country. Therefore, for GRB to be meaningful, it must necessarily begin with purposive gender planning for each scheme/sector—'first by identifying the gender gaps in the sector and then delineating prioritised actions points to address the gender gaps' (Brody, 2009: 54). Indeed, even within this framework, to advance much further greater understanding and capacity of what can be achieved with GRB are first required.

1.2.3 Gender, WASH and Climate Change

Climate variability and change is fast becoming a contextual case in the analysis of gender relations. Although climate change affects everyone, its drivers and effects are not gender neutral. Climate change magnifies existing inequalities, reinforcing the disparity between women and men in their vulnerability to and capability to cope with changing climatic and environmental conditions (Dankelmann, 2010). Recognising the importance of the subject, the 52nd session of the Commission on the Status of Women (2008) identified gender and climate change as its key emerging issue. Specific provision on Financing for Gender Equality and the Empowerment of Women urged governments to:

> Integrate a gender perspective in the design, implementation, monitoring, and evaluation and reporting of national environmental policies, strengthen mechanisms and provide adequate resources to ensure women's full and equal participation in decision making at all levels on environmental issues, in particular on strategies related to climate change and the lives of women and girls. (Hemmati, 2008)—Resolution 21(jj) (E/CN.6/2008/L.8)

However, a pertinent question is—how is gender and climate change issue related to WASH? IDS (2008) documents the impact of climate change on water supply and concludes that climate change impact will lead to increasing frequency and intensity of floods and deteriorating water quality. This is likely to have a particularly harsh effect on women and girls because of their distinct roles in relation to water use and their specific vulnerabilities in the context of disasters. Extreme events will bring in more uncertainty for water availability leading to increased water stress for women. Similarly, heavy rainfall leading to flood will accentuate sanitation condition and water contamination leading to unhygienic conditions, which will affect women more than men because of the vulnerability. It will also increase women's workloads, as they will have to devote more time to collecting water and to cleaning and maintaining their houses after flooding. That time could be spent in school, earning an income or participating in public life. Though the chapters in this volume do not specifically contextualise gender and climate change issues, it has been brought as a challenge for future WASH programme, and therefore should be considered as part of planning and implementation of WASH programme.

1.3 About the Book

This book focuses on ways to achieve desired gender outcomes through water and sanitation programmes in India rather than reproducing the essential tenets of the water sector. The recent study 'Diverting the Flow' is a seminal volume to help fill the knowledge gap in the water sector (Zwarteveen et al., 2013). Our collection of essays contributes to current debates in the water and sanitation sectors by further reinforcing the need for process, resources, capacity and evaluation. It shows through special attention to gender and to case studies of successful implementation that reaching out to the unreached is possible. This is conditional to a concerted attempt to advancing equal access to gender-responsive basic water and sanitation, increasing women's voice in decision-making and investment in their leadership and supporting their efforts in bringing a change process. Process is critical in this and needs further development in the Indian water and sanitation context.

The book is divided into four sections. Section 1 provides a conceptual overview of gender in the water and sanitation sector in India. Chapters in this section focus on exploring ways to incorporate gender dimensions in water management and the broader water and sanitation agendas in India. Section 2 provides case studies involving women in the water sector and discusses women's participation, roles and voices. Section 3 focuses on the cases of women's participation in the sanitation sector with the focus on innovative ways in which women's role and participation can be upscaled.

1.3.1 Section 1

In India, women play pivotal roles in the water and sanitation sector. Although policy makers in this sector have started realising the importance of women's role in ensuring the success of the project and sustainability of the resources in the water sector, there is still little common understanding of how gender issues can be addressed and how women's participation in planning, designing and implementation can be upscaled. Chapter 1 by Prakash, Cronin and Mehta focuses on exploring ways in which these gender dimensions can be optimally included in the water sector.

Chapter 2 by Lala, Basu, Jyotsna and Cronin reviews the existing conceptual frameworks and current practices to strengthen the implementation of WASH programmes in India. They review a series of gender frameworks and critique timelines and processes required for strengthening gender outcomes in the WASH context. On the basis of an e-query run on both the Solution Exchange (Water and Gender Communities of Practice), the chapter broadly categorises the steps in designing WASH programmes to adequately include gender and equity, with specified timelines. The chapter also presents potential indicators to measure the effectiveness of each phase.

Chapter 3 by Kabir, Vedantam and Kumar assesses the vulnerability of rural households due to lack of sufficient water for domestic and productive needs and shows how multi-use approaches to water can help. Given the wide range of factors—natural, physical, social, human, economic, etc.—influencing this vulnerability, the authors develop an index that helps communities to compute the vulnerability of a household to problems associated with lack of water for domestic and other productive needs. The proposed index has six sub-indices, in the areas of water supply and use; family occupation and social profile; social institutions and ingenuity; climate and drought proneness; water resources availability; and financial stability. Using case studies from three regions of Maharashtra, the authors argue that the identification of vulnerable households can help in devising systems to reduce the hardships faced by women. The computed index can be used for designing of water supply augmentation schemes and retrofitting of existing water supply infrastructure to meet the multiple water needs of poor rural households with special focus on women.

Chapter 4 by Prakash and Goodrich documents the approaches, outputs and outcomes of a unique initiative called 'Crossing Boundaries' (CB). The project focused on education, impact-oriented research, networking and advocacy as a combined effort to contribute to a paradigm shift in water resources management in four South Asian countries. The project imparted interdisciplinary approach with gender and equity concern emphasised into technical education through curriculum change, guiding student's thesis and training faculties in dealing with interdisciplinary water issues. The emphasis of this project was on long-term educational inputs for shaping attitudes and perceptions, and teaching the skills of interdisciplinarity, comprehensive analysis and intervention. The project was implemented by a group of institutions with proven interest and track record in integrated, interdisciplinary and gender-sensitive approaches to water resources management and was completed in 2011. This chapter documents the regional, collaborative, partnership-based capacity-building initiatives undertaken by the project.

Chapter 5 of Section 1, by Sinha, reviews the current capacity-building initiatives of WASH in India from a gender and equity lens. The chapter reviews the curriculum of select Communication and Capacity Development Unit (CCDU) and NGO programmes to understand how well gender and equity are placed in the overall programming context and where bottlenecks that require action, within a framework for action, are located. The chapter suggests that the current training programmes of the Government of India are lacking a concerted approach to gender and equity in WASH. The training curricula lack focus on gender issues and the recipients of this training are largely men. The author raises several critical questions on trainings.

1.3.2 Section 2

Section 2 examines successful case studies involving women. It has six chapters that discuss women's participation, roles and voices in the water management sector in India. Chapter 6 by Wani, Anantha and Sreedevi identifies critical factors essential for enhancing gender participation through watershed programmes in the semi-arid regions of India. It details several drivers of success for improving benefits to women and vulnerable groups. The authors argue that watershed programmes should look beyond land development activities to address rural livelihoods to make substantial differences in the socio-economic status of women and vulnerable groups. Further, implementing and coordinating agencies need to recognise the pressing need for both resources and policy support to address gender equity.

In Chapter 7, Bastola attempts to understand women's participation in decentralised water institutions established by the Jalswarajya Project in the state of Maharashtra. The author examines the attempts to challenge the power structures restricting women to play a stronger role in community decision-making processes. The chapter highlights that rural elites have used gender and caste-based discrimination within the Jalswarajya Project to perpetuate existing power bases in decentralised institutions. The author suggests that good water governance, when promoted without convergence and addressing existing social dynamics, can confound hopes of change for women, especially among poor marginalised populations.

The chapter by Mehta and Saxena (Chapter 8) addresses questions related to the extent and type of women's participation in water-related issues and their implications in Mewat, a water-scarce region where availability and quality of water are vital concerns for survival. The chapter highlights the role of water in perpetuating poverty in the region in the state of Maharashtra. The author highlights differing gender water priorities in which men stress the distance of water source while women give priority to water quality of the source. This results in women fetching water from closer sources but potentially more contaminated sources, contributing to high incidence of waterborne diseases with its associated economic implications. The authors call for a more active role of women in the decision-making process and involvement in the construction, maintenance and usage of water sources.

Chapter 9 by Chakma, Medeazza, Singh and Meshram outlines the impact of fluoride on tribal women in selected parts of Madhya Pradesh and exemplifies interventions involving women, which have led to improvements in the nutritional and health situation. The chapter postulates that even though excessive fluoride levels in water affect the entire population, the impact is more profound on women. This is due to the lower nutritional levels of women, their reproductive function and the higher burden of household chores they bear, including fetching water. Assessing the impact of the Regional Medical Research Centre for Tribals' work demonstrates how an integrated fluoride mitigation approach linking safe drinking water with nutritional supplementation when led by women can help reduce the prevalence of fluorosis, both among women and among the population at large.

Chapter 10 by Mani, Rao, Reddy and Babu details the different processes adopted in mainstreaming gender for improved water use efficiency. It deliberates on the policy matters related to role of women in natural resource management. Recognising that water resources planning, development and management have always been handled by a closed group of experts, the authors highlight how this has led to the alienation of the poor and the weak. The latter are compelled to use groundwater resources for irrigation. The chapter presents the case of Andhra Pradesh Farmer Managed Groundwater System (APFAMGS) project, which has successfully implemented an approach to the just usage of available water resources in seven drought-prone districts in the state based on women's empowerment. The authors suggest that this approach has inspired other projects to relook gender approaches beyond simple involvement of men and women, to influencing the overall attitude towards women and, most importantly, respecting their technical and analytical skills.

Chapter 11 by Prasad, Acharya and Basu reviews the implementation of a drinking water supply systems project in the eastern state of Jharkhand from the gender lens. This project was rolled out through a government–NGO partnership whereby women's SHGs were centrally involved in planning and implementation. Three critical factors that characterise the involvement of women in the drinking water sector are put forward involving the key role of women in leadership, women's central importance in participation and finally getting women as users. It emerges clearly that in water supply schemes, the involvement of women is mandatory in order to ensure sustainability. The authors argue that in order to make the woman's participation efficient, she needs to be adequately empowered against discrimination or vulnerability. The woman–water linkage manifests a critical scenario in gender deprivation if water is not readily available to the household. Availability of water at home addresses issues of women's well-being and progression. Hence, SHGs should consciously promote water issues as part of the health and economic programmes.

1.3.3 Section 3

In the third section, we explore experiences from developing and middle-income countries that can help improve understanding of the gendered approaches, especially in relation to women's participation in the sanitation sector. Chapter 12 by Kale and Zade focuses on dealing with the issues of gender role and participation through an innovative approach developed by Watershed Organisation Trust (WOTR) in the state of Maharashtra. The authors argue that for effective participation of women, their capacities need to be strengthened by organising them in village-level organisations and giving them the required space to participate. If appropriate and adequate opportunities and institutional spaces are created for women with a sound capacity-building strategy and financial autonomy, then domestic water availability and accessibility, health and sanitation may be effectively addressed. Though persuading men to accept women's leadership at institutional level is quite challenging, trainings on gender sensitivity with follow-up meetings (for men and women) has proven to be an effective strategy. The authors argue that for meaningful inclusion and participation of women in watershed development, there is a strong need to address their concerns on domestic water, health and sanitation. It is clear that separate spaces in the form of sub-institutions and sub-committees (such as village health committee) with financial autonomy and well-defined decision-making process are needed.

Chapter 13 by Medeazza, Jain, Tiwari, Shukla and Kumar presents lessons from the implementation of community-led total sanitation (CLTS) in Madhya Pradesh and shows how women play a decisive role in rendering their community Open Defecation Free (ODF) under the state-level sanitation campaign. The chapter draws key recommendations from selected case studies of women-led sanitation interventions. To achieve ODF communities, the authors argue that there is a need for greater emphasis on demand generation for toilet *use*. *Additionally*, sequencing of fund release to communities is key: allocations should be given to communities only after communities have reached ODF status (i.e. as post-incentive, not pre-subsidy). Collective behaviour change can be accelerated through facilitating ownership of the communities and a desire for dignity and self-respect. The monitoring mechanism of toilets *constructed* versus toilets *used* and ODF status must be strengthened. This chapter highlights that over a relatively short period of time, CLTS can be a potentially successful approach to render communities ODF and that in the process women have emerged as empowered natural leaders, whose roles within their communities can go way beyond the initial sanitation objectives.

Chapter 14 by Saxena, Mujumdar and Medeazza discusses a new and innovative approach—tailored towards training women to provide inclusive and women-centric WASH services. The chapter draws from the learning of 80 camps at *Gram Panchayat* level (the smallest unit of local government in India) in Madhya Pradesh. This approach is based on 'on-site' principles, which keep women at the core of the entire process. Through innovative training approaches that revolve around women's commitments, the strategy empowers women to take a more prominent leadership role through playful awareness exercises. The chapter also presents the findings from field testing of these camps and underscores the enormous improvement in message retention, behavioural change, involvement of women in planning of village WASH-related components as well as child participation in school sanitation.

Chapter 15 by Mehrotra and Singh examines the role of ASHA (Accredited Social Health Activist) in accelerating sanitation in the state of Uttar Pradesh. While many of the other chapters examine NGO-led processes, this chapter deals with a Government of India-led sanitation initiative. This work was carried out by the frontline functionary of the Government Health system, the ASHA workers who normally provide very basic health support at village level. In this project, over 13,000 ASHAs have been trained on WASH behaviours in eight districts of Uttar Pradesh. The authors describe that the trained ASHAs are now a crusader of sanitation issues and have not only started using toilets themselves but also started promoting its use among the community members by linking open defecation's ill-effect with safe health.

1.3.4 Section 4

The concluding chapter of this volume (Chapter 16) pulls together the central themes of the case studies where we offer new insights on gender inclusion in WASH and outline key action agendas that can take this dialogue forward in India.

Disclaimer

The views expressed herein are those of the authors and do not necessarily reflect the views of UNICEF, the United Nations, SaciWATERs or S M Sehgal Foundation.

Notes

1. Quintile is a method of dividing the entire population of the country into five equal parts, based on a wealth index derived from 33 family assets under the National Family Health Survey (NFHS).
2. Using monthly per capita consumer expenditure (MPCE).

References

Ahmed, Sara (ed). 2005. *Flowing Upstream: Empowering Women through Water Management Initiative in India*. New Delhi: Foundation Book.

Baby V. Kurian and V. Ratna Reddy. 2013. How effective are the new WASH security guidelines for India? An empirical case study of Andhra Pradesh. *Water Policy*, 15(4): 535–553.

Brody, Alyson. 2009. *Gender and Governance: Overview Report. Bridge–Development–Gender*. Sussex, UK: Institute of Development Studies.

Cap-Net, India, SOPPECOM, Women Water Network (WWN). 2007. *Water for Livelihoods: A Gender Perspective*. (Based on writings of S. Ahmed, S. Arya, K.J. Joy, S. Kulkarni, S.M. Panda and S. Paranjape) Ch. 5, pp. 60–76. Pune: Cap-Net, India, SOPPECOM and WWN.

Chand, D., C. Dey, M. Prakash and M. Kullappa. 2004. 'Review Report on Total Sanitation Campaign/Swajaldhara Projects In Uttar Pradesh'. Available at http://ddws.nic.in/popups/rev_up_04.pdf (accessed on 8 March 2011).

Cleaver, Frances. 1998. Choice, complexity and change: Gendered livelihoods and the management of water. *Agriculture and Human Values*, 15(4): 293–299.

Coles, A. and T. Wallace (eds). 2005. *Gender, Water and Development*. Oxford: Berg Publishers.

Cronin, Aidan. 2013. Impact of Poor WASH on Child Health and Ability, Presentation to the WaterAid Workshop on WASH and Disability, New Delhi, January 2013.

Cullet, Philippe. 2009. New policy framework for rural drinking water supply: Swajaldhara guidelines. *Economic and Political Weekly*, 44: 47–54.

Dankelmann, Irene. 2010. *Gender and Climate Change: An Introduction*. London, UK: Earthscan.

Esrey, S.A. 1996. Water, waste, and well-being: A multi-country study. *American Journal of Epidemiology*, 143(6): 608–623.

Fewtrell, L., A. Prüss-Üstün, R. Bos, F. Gore and J. Bartram. 2007. *Water, Sanitation and Hygiene: Quantifying the Health Impact at National and Local Levels in Countries with Incomplete Water Supply and Sanitation Coverage*. World Health Organization, Geneva, 2007 (WHO Environmental Burden of Disease Series No. 15).

GLAAS. 2012. UN-Water Global Annual Assessment of Sanitation and Drinking-water report: The challenge of extending and sustaining services. World Health Organization, Geneva, Switzerland. Available at http://www.unwater.org/downloads/UN-Water_GLAAS_2012_Report.pdf (accessed on 2 May 2013).

Hemmati, M. 2008. *Gender perspectives on climate change*. Interactive expert panel on the theme. UN Commission on the Status of Women. 52nd session, 25 February–7 March 2008. New York.

IDS. 2008. *Gender and Climate Change: Mapping the Linkages*. *Bridge–Development–Gender*. Sussex, UK: Institute of Development Studies.

Jacob, Nitya, Shubhangi Sharma, Sunetra Lala and Sudakshina Mallick. 2009. *Consolidated Reply: Impact of School Sanitation on Adolescent Girls—Experiences; Examples*. New Delhi: United Nations Solution Exchange India.

Joshi, D. 2005. Misunderstanding gender in water: Addressing or reproducing exclusion, in A. Coles and T. Wallace (eds), *Gender, Water and Development*, Ch 8. Oxford: Berg Publishers.

Joshi, D. 2011. Caste, gender and the rhetoric of reform in India's drinking water sector. *Economic and Political Weekly*, xlvi(18): 56–63.

Kumar, Ganesh S., Sitanshu Sekhar Kar and Animesh Jain. 2011. Health and environmental sanitation in India: Issues for prioritizing control strategies. *Indian Journal of Occupational Environmental Medicine*, 15(3, Sep–Dec): 93–96.

Lahiri-Dutt, Kuntala (ed). 2006. *Fluid Bonds: Views on Gender and Water*. Calcutta: Stree.

Lahiri-Dutt, Kuntala and Robert Wasson (eds). 2008. *Water First: Issues and Challenges for Nations and Communities*. New Delhi: SAGE Publications.

MDWS. 2011. *MDWS Strategic Plan for Water and Sanitation, 2011–22*. Ministry of Drinking Water and Sanitation, Government of India.

Meinzen-Dick, R.S. and M. Zwarteveen. 1998. Gendered participation in water management: Issues and illustrations from water users associations in South Asia. *Agriculture and Human Values*, 15(4): 337–345.

Mishra, Yamini and Nivedita Sinha. 2012. Gender responsive budgeting in India: What has gone wrong? *Economic and Political Weekly*, xlvii(17): 50–57.

Moser, C. 1993. *Gender Planning and Development: Theory, Practice and Training*. New York: Routledge.

Narain, V. 2003. Water scarcity and institutional adaptation: Lessons from four case studies. in *TERI, Environmental Threats, Vulnerability, and Adaptation: Case*

Studies from India, pp. 107–120. The New Delhi: The Energy and Resources Institute.
NFHS. 2006. *National Family Health Survey 3*. Government of India, New Delhi.
NSSO. 2010. National Sample Survey Office, Housing Condition and Amenities in India 2008/09, 65th Round, released in November 2010.
O'Reilly, Kathleen. 2006. "Traditional" women, "modern" water: Linking gender and commodification in Rajasthan, India. *Geoforum*, 37(6): 958–972.
Pattanayak, Subhrendu K., Christine Poulos, Jui-Chen Yang and Sumeet Patil. 2010. How valuable are environmental health interventions? Evaluation of water and sanitation programmes in India. *Bull World Health Organ* 88(7). Sourced at http://www.scielosp.org/scielo.php?pid=S0042-96862010000700013&script=sci_arttext (accessed on 12 June 2013).
Planning Commission. 2011. *Faster, Sustainable and More Inclusive Growth: An Approach to the Twelfth Five Year Plan (2012–17)*. Government of India.
Prakash, Anjal. 2012. Groundwater vending and appropriation of women's labour: Gender, water scarcity and agrarian change in a Gujarati Village, India, in Zwarteveen, Margreet, Sara Ahmed and Suman Rimal Gautam (eds), *Diverting the Flow: Gender Equity and Water in South Asia*, pp. 312–336. New Delhi: Zubaan.
Prakash, Anjal and R.K. Sama. 2008. *Social Undercurrents in a Gujarat Village: Irrigation for the Rich vs. Drinking Water for the Poor. Water Conflicts in India: A Million Revolts in the Making*, pp. 38–43. New Delhi: Rutledge.
Prakash Anjal, Sreoshi Singh, Chanda Gurung Goodrich and S. Janakarajan. 2012. Introduction: An agenda for pluralistic and integrated framework for water policies in South Asia, in Prakash Anjal, Sreoshi Singh, Chanda Gurung Goodrich and S. Janakarajan (eds), *Water Resources Policy in South Asia*. New Delhi: Rutledge.
Rao, Smriti. 2008. Reforms with a female face: Gender, liberalization, and economic policy in Andhra Pradesh, India. *World Development*, 36(7): 1213–1232.
Sangameswaran, Priya. 2010. Rural drinking water reforms in Maharashtra: The role of Neo-liberalism. *Economic and Political Weekly*, 55(4): 62–69.
Seager, J. 2010. Gender and water: Good rhetoric, but it doesn't 'count', editorial. *Geoforum*, 41: 1–3.
Shah, Amita and Seema Kulkarni. 2013. Interface between water, poverty and gender empowerment: Revisiting theories, policies and practices, in Anjal Prakash, Sreoshi Singh, Chanda Gurung Goodrich and S. Jankarajan (eds), *Water Resources Policies in South Asia*, pp. 19–36. New Delhi: Routledge.
Starkl, Markus, Norbert Brunner and Thor-Axel Stenström. 2013. Why do water and sanitation systems for the poor still fail? Policy analysis in economically advanced developing countries. *Environmental Science and Technology*, 47(12): 6102–6110.
Times of India. 2013. Rapists on prowl in loo-less Bihar, Jan 2013, Patna Edition.
UNICEF. 2011. Promoting gender equality through UNICEF-supported programming in young child survival and development—operational guidance.

UNICEF. Sourced at http://www.unicef.org/gender/files/Survival_Layout_Web.pdf (accessed on 12 July 2013).

UNICEF. 2012a. Equity in Drinking Water & Sanitation in India, Perspectives on Equity and Gender in the WASH Sector in India; UNICEF India 2012, 6 pages.

———. 2012b. 2001 and 2011 Census WASH access map, UNICEF India WASH Section. New Delhi.

———. 2012c. *The State of the World's Children 2012: Children in an Urban World*. New York, UNICEF.

UNICEF, FAO and SaciWATERs. 2013. Water in India: Situation and Prospects. 105 pages. Available at http://www.unicef.org/india/media_8098.htm (accessed on 23 April 2013).

UNICEF and WHO. 2012. Progress on Drinking Water and Sanitation—2012 update. Joint Monitoring Program on Water supply and Sanitation. New York, USA. Available at http://www.wssinfo.org/fileadmin/user_upload/resources/JMP-report-2012-en.pdf (accessed on 2 May 2013). Page 30.

United Nations. 2010. *The World's Women: Trends and Statistics*. Department of Economic and Social Affairs, United Nations Statistics Division, New York, USA.

UN Water. 2006. Gender, Water and Sanitation: A Policy Brief. Inter-agency Task Force on Gender and Water (GWTF). UN-Water and the Interagency Network on Women and Gender Equality (IANWGE). Available at http://www.un.org/waterforlifedecade/pdf/un_water_policy_brief_2_gender.pdf (accessed on 12 August 2013).

WaterAid. 2011. Including disabled people in sanitation and hygiene services: WaterAid and Share briefing note June 2011, 8 pages.

———. 2013. Mainstreaming disability and ageing in water, sanitation and hygiene programmes. WaterAid and WEDEC. WaterAid. UK.

Zwarteveen, M. 2008. Men, masculinities and water: Powers in irrigation. *Water Alternatives*, 1(1): 111–130.

———. 2011. Questioning masculinities in water. *Economic and Political Weekly*, xlvi(18): 40–48.

Zwarteveen, M., S. Ahmed and S. Gautam (eds). 2012. *Diverting the Flow: Gender Equity and Water in South Asia*. New Delhi: Zubaan Publishers.

SECTION 1

Conceptual Underpinning

2

Accelerating Gender Outcomes: The WASH Sector

Sunetra Lala, Malika Basu, Jyotsna and Aidan A. Cronin

2.1 WASH in India

Safe, adequate and sustainable water, sanitation and hygiene (WASH) for all is one of the main goals repeatedly emphasised by the global community (Scanion et al., 2004). Unfortunately, 11 per cent of the world's population (783 million people) still lack access to clean water, 2.5 billion people continue to do without improved sanitation, and millions die each year due to water-related diseases (WHO/UNICEF, 2006). As of 2010, 97 million people in India did not have access to improved water supplies, and 626 million practised open defecation (WHO/UNICEF, 2012).

Amongst WASH stakeholders, one of the largest groups is identified by gender (Jiménez and Asano, 2008). In 72 per cent households worldwide, women and girls are responsible for collecting water (UNICEF, 2011). The Government of India has also recognised the importance of gender. The Planning Commission has stated that women and children who constitute 70 per cent of the population deserve special attention; thus, ending gender-based inequities faced by girls and women must be accorded the highest priority (Planning Commission, 2011). India's national rural drinking water and sanitation programmes are also aimed at improving service provision and explicitly state the need for social and gender equity (MDWS, 2011). However, programmes mostly promote a universal access approach, rather than a gendered approach. This has made the differential needs of men and women invisible (Lala and Basu, 2012). The existing approaches to WASH programming in India are often ghettoised, focusing mostly on technical improvements and sectorial responses. These do not pay adequate attention to social and sustainability goals, including critical issues around community engagement (Cronin and Burgers, 2011).

This chapter identifies two key gaps to help the WASH-gender discourse in India to move ahead. First, we examine issues of gender equity and the need for a gender analysis to understand how men and women undertake planning, decision-making and implementation of WASH programmes. Second, we review existing gender frameworks to understand how these capture gender dynamics in WASH. Finally, we apply the findings to the desk review of good practices with the objective of strengthening gender-sensitive planning and programming in the WASH sector.

2.2 Gender and WASH in India

Community-managed WASH service delivery is taking root in some parts of India as the sustainable and equitable way forward, and moving away from relying only on capital-intensive WASH projects. Examples of decentralised community-driven water harvesting projects are visible in the arid regions of Rajasthan and Gujarat (Agarwal et al., 2001). However, a community is composed of diverse groups and individuals (commanding different levels of power, impact and capability to express their needs and rights). Hierarchies influence water distribution and access in India. Patriarchal patterns also influence decision-making at the community level leading to women being excluded from planning, decision-making and implementing WASH.

In India, water management is a part of the informal economy, ruled by traditional customs. Women often bear the brunt of lack of adequate water and sanitation facilities. On the one hand, once water management enters the public domain, it is subject to social norms ensuring men remain managers and women adhere to decisions made for them. On the other hand, women are also overburdened by household work and economic efforts (Agarwal, 1994). Additionally, water rights are closely tied to land tenure provisions and are linked to land issues. Patriarchal relations also express themselves in control over land rights (Agarwal, 1994; Lerner, 1944). Furthermore, women do not have any control over the benefits of their water-related labour as they work on land allotted to their husbands or fathers (Zwarteveen et al., 2012).

The provision of water for the realisation of essential needs has always been a woman's responsibility. In several parts of India, women spend hours fetching water from distant sources daily. This wields tremendous physical burden on women and impacts their health. The condition of pregnant women is worse. Poor water quality and inadequate sanitation negatively impact their food security, levels of nutrition and livelihood choices. It also significantly reduces time available for other development prospects, such as education. The economic value of this unpaid role is huge; in India, it is estimated that women fetching water spend 150 million workdays per year, equivalent to a loss of 10 billion rupees to the country (UN ESCAP, 2005).

Lack of proper sanitation facilities also leads to high dropout rate of girls from schools, especially around the onset of puberty. This, in turn, affects their health and exposes them to the risk of assault when they seek for privacy in secluded spots, away from the school, to relieve themselves. However, even when sanitation facilities are available, there are challenges involved in ensuring all students have equal access. For instance, children from Dalit families are often discriminated against, when accessing common toilets (Jacob et al., 2009).

Although it is accepted that the involvement of both men and women is required for sustainable WASH services, women often miss out on extension services and job opportunities in the WASH sector. In many cases, water and sanitation projects involve construction and then operation and maintenance (O&M) work. These are not regarded as activities for women and they are not offered the training necessary to equip them to effectively participate in these fields.

To add to the existing complexities, there are no uniform gender-disaggregated data available in the WASH sector (UNICEF, FAO and SaciWATERs, 2013). This lack of data influences our understanding of the relationship between gender and WASH services (Seager et al., 2009). In the WASH sector, gender differences carry important implications (Zwarteveen et al., 2012). Given the gender inequalities prevailing in India, it is essential for WASH programmes to take into account how they might benefit women. For example, women's participation could entail access to information, decision-making and control of water resources.

The recently growing consensus on the need to involve both men and women equally in WASH programmes comes from the recognition that WASH services can become more efficient, user-focused, financially viable and sustainable with the active involvement of men and women (Zwarteveen et al., 2012). A generic people-centric approach does not inevitably ensure that the needs and interests of women and men are adequately reflected in WASH programming. Gender analysis helps to redress inequalities and inequities, examines barriers to participation, and envisages potential outcomes of development interventions to ensure that women contribute to and benefit from WASH programmes.

2.3 Gender Analysis for WASH Programming

Gender analysis is a set of processes used for understanding and assessing differences in the lives of women and men, their participation in social and economic life, and their differential responses and impacts to policies, programmes and services. In the WASH context, gender analysis helps acknowledge disadvantages that women and girls face by drawing attention to the interface between institutional, socio-economic and cultural systems. It assesses their specific needs and the likely impacts of policies, programmes and services. This facilitates the articulation of their viewpoints, making their inputs a critical part of WASH interventions. It also helps in ensuring that their needs and issues are clearly identified and addressed through each step of planning, implementation, evaluation and management. Thus, women and girls' equitable engagement is promoted in the process (March et al., 1999).

Gender analysis, a part of wider situational analysis, required for every programme (Zwarteveen et al., 2012), considers the experiences and responsibilities of women, taking into account their relatively lower levels of access to resources and decision-making processes. The analysis promotes the understanding that treating everyone in the same way helps make some things more equal, but it is insufficient to meet specific needs of women across caste, class, age, disability, marital status, etc. Gender analysis for WASH programming helps to ensure that the roles and responsibilities of men and women complement each other to yield the best results (Derbyshire, 2002). It also enables both genders to contribute to make water and sanitation schemes work better. It informs the diverse interlinked processes needed to assess the differential impacts on women and men, girls, and boys (Rennie and Singh, 1996).

Though a number of frameworks exist to help analyse gender in development programmes, a dedicated framework to capture gender inequities in

WASH specifically for India is currently missing. In this chapter, we develop a structured approach to gender analysis by looking across the typical phases of a WASH programme. Corresponding indicators were designed to help monitor progress towards positive gender and equity outcomes.

2.4 Methodology

We analysed seven well-known gender analysis frameworks used in development research and planning with proven values for assessing and promoting gender issues (March et al., 1999). Previous analyses, such as the gender equality framework (USAID, 2008) and the performance assessment framework (AusAID, 2005) have also been based on these but are excluded from the analysis, as they are not specific to the Indian WASH setting. The need to further interpret and map these frameworks to assess their potential value for India, therefore, remains. To inform the framework review and to link it with the practical implementation of WASH in India, a discussion was initiated with grassroots practitioners, via Solution Exchange. Solution Exchange is an email-based knowledge-sharing platform developed and managed by the United Nations system in India since 2005. It comprises 10 Communities of Practice (CoPs) with over 35,000 members sharing and exchanging experiences and knowledge, and collectively solving development problems. The Water and Gender CoPs, 2 of the 10 CoPs, jointly hosted an e-discussion on equity in WASH Programming in April 2012. The following questions were addressed:

- What are the existing structured approaches for ensuring inclusion and equity in WASH programmes?
- What are the essential phases that have been identified under such an approach/framework? What is the phased timeline if any, and phased break-up of resources? In addition, what are the essential quality checks to be kept in mind at each phase?
- What are the emerging indicators that can facilitate effective monitoring of inclusion and equity in WASH?
- What could be the best practices in the Indian water and sanitation sector that can illustrate a structured and consistent approach to gender and equity outcomes?

2.5 Frameworks for Gender Analysis in Development Practice—A Review

The seven gender frameworks reviewed include the Harvard Analytical Framework (Overholt et al., 1985); People-Oriented Planning (POP) Framework (Anderson et al., 1992); the Moser Framework (Moser, 1993); Gender Analysis Matrix (GAM) (Parker, 1993); Capacities and Vulnerabilities Analysis Framework (Anderson and Woodrow, 1989); Women's Empowerment Framework (Longwe, 1991); and the Social Relations Framework (Kabeer, 1985). Six broad categories were identified in WASH programming—participation, access to services, control over resources, benefits to women, governance, and O&M. Sub-components of each category were identified, and a cross-comparison was carried out against the seven frameworks (Table 2.1).

Table 2.1:
Gender framework: WASH project component comparative analysis

	Frameworks						
Category	*Harvard Analytical*	*People-Oriented Planning (POP)*	*Moser*	*Gender Analysis Matrix (GAM)*	*Capacities and Vulnerabilities Analysis*	*Women's Empowerment (Longwe)*	*Social Relations Approach*
1 Participation							
1.1 Quality of process (i.e. gender balance in participation)	×	×	✓	×	×	×	✓
1.2 Quality (representation resulting in gender-sensitive decisions)	×	×	×	✓	×	×	✓
1.3 Networks (women's access to formal or informal networks)	×	✓	✓	×	✓	×	✓
1.4 Inclusion (demographic composition of population)	×	✓	×	✓	✓	✓	✓

(Table 2.1 Continued)

(Table 2.1 Continued)

Category	Harvard Analytical	People-Oriented Planning (POP)	Moser	Gender Analysis Matrix (GAM)	Capacities and Vulnerabilities Analysis	Women's Empowerment (Longwe)	Social Relations Approach
	Frameworks						
1.5 Community power dynamics (in relation to gender)	✓	×	×	✓	✓	✓	✓
2 Access to Services							
2.1 Hardware (toilets, taps)	✓	✓	✓	×	×	✓	✓
2.2 Coverage (extent to which people are reached)	×	✓	×	×	✓	×	×
2.3 Distance (time to access services)	×	✓	×	×	✓	×	×
3 Control Over							
3.1 Source	✓	✓	✓	✓	✓	✓	✓
3.2 Land (where source rights belong to land owner)	✓	✓	✓	✓	✓	✓	✓
3.3 Household decision-making (on procurement, management and distribution in household)	×	×	✓	✓	×	✓	✓
4 Benefits to Women							
4.1 Livelihood	✓	✓	✓	✓	×	✓	✓
4.2 Health	✓	✓	×	×	×	✓	✓
4.3 Education	✓	✓	✓	×	×	✓	✓
4.4 Time (how much time is saved, more work and less people involved)	×	✓	×	✓	✓	×	×
5 Governance							
5.1 Institutional strengthening in terms of gender (*Gram Panchayat* and Public Health Engineering Department)	✓	✓	✓	×	✓	✓	✓

(Table 2.1 Continued)

(Table 2.1 Continued)

	Category	Harvard Analytical	People-Oriented Planning (POP)	Moser	Gender Analysis Matrix (GAM)	Capacities and Vulnerabilities Analysis	Women's Empowerment (Longwe)	Social Relations Approach
		Frameworks						
5.2	Voice (gender empowerment, especially with reference to decision-making)	×	✓	×	✓	×	✓	✓
6 Operation and Maintenance and Management								
6.1	Participation	×	✓	×	×	×	✓	✓
6.2	Contribution (in terms of time, labour and/or money)	×	×	×	×	✓	×	✓
6.3	Monitoring	×	×	×	×	✓	×	×

Many other gender analysis frameworks exist, such as the Good Practices Framework (CARE, 2012) and the Gender Analysis Tool (Status of Women Canada, 2013). The Gender Analysis Tool (Goldberg and Lang, 2004) was formulated to comprehend gender analysis in poverty reduction to 'develop a clear set of policy recommendations for poverty reduction in its community'. Another valuable tool is the gender-focused field diagnostic kit (Bishop-Sambrook, 2000), based on the entire programme cycle. These are cited as examples of other approaches that draw upon the elements of the seven chosen frameworks.

Each framework used focuses on factors that influence or perpetuate gender differences. For instance, the Harvard Analytical Framework focuses on the 'socio-economic activity profile' of who does what, when, where and for how long. It also focuses on the 'access and control profile', that is, who has access and control over resources and benefits. In addition, it focuses on the 'influencing factors' (i.e. what enables/facilitates the socio-economic profile as well as access and control profile). The POP Framework, on the other hand, is based on three major components—determinants analysis, activity analysis and use and control of resources analysis—geared towards understanding equal distribution of goods and services. It focuses on who, where and when, including how resources are utilised. It emphasises on control and benefits. The Moser Framework uses two main tools for gender analysis. Firstly, this includes identification of gender roles in terms of women's roles around production, reproduction and community. Secondly, it incorporates gender needs (or interests met) assessment in terms of practical gender needs and strategic gender interests. In this regard, the framework primarily captures the control over resources and touches briefly on benefits and participation. The GAM uses participatory methodology to facilitate the definition and analysis of gender issues by affected communities. Each project objective is analysed at four societal levels—women, men, household and community—by various groups of stakeholders. It carries out analysis by discussing each project objective and its impacts on men's and women's labour practices, time, resources and other socio-cultural factors, such as changes in social roles and status.

Capacities and Vulnerabilities Analysis Framework is also more inclusive of different WASH indicators by including 'control over resources' and some 'governance' aspects. However, it does not pay sufficient attention to benefits. The Women's Empowerment Framework puts forward five levels of equality, namely, control, participation, conscientisation (attaining equal understanding of gender roles and a gender division of labour that is fair and agreeable), access (equal access to the factors of production by removing discriminatory provisions in laws) and welfare (having equal access to material welfare such as food, income, medical care). The Women's Empowerment Framework, like others, captures control over resources and resultant benefits. Importantly, it captures governance issues as well. Finally, the Social Relations Framework analyses existing gender inequalities on the distribution of resources, responsibilities and power-relationships between people, and how their relationship to resources and activities are reworked through institutions. It emphasises human well-being as the final goal of development. The Social Relations Framework is the only comprehensive framework that takes social relations and power dynamics into account and shows that these play a significant role in highlighting gender aspects in WASH.

2.6 Drawing from the Existing Frameworks: A Hybrid Framework for WASH

None of the frameworks reviewed capture all components that comprehensively analyse the gender inclusiveness of WASH programmes (Table 2.1). The Harvard Analytical Framework is weaker on participation and governance indicators. The POP Framework does not take quality of process, quality of representation, O&M and management into account. The GAM lacks indicators on access to services, O&M and management—although it does have some indicators relating to participation and benefits. The Capacities and Vulnerabilities Analysis Framework does not pay sufficient attention to benefits. Similarly, the Moser Framework partially covers benefits, but is weak on important indicators such as access and participation. In general, the frameworks are weak on O&M and do not give prominence to participation, a key indicator in WASH that makes the programme gender inclusive. Further, intra-household decision-making such as who fetches, stores, utilises and manages water for domestic activities, an important unit of enquiry, is reflected only under Moser, GAM and the Social Relations Framework.

Each framework was developed at a particular time keeping specific development objectives/programmes in mind. Thus, they do not necessarily reflect or capture the gender dynamics of a WASH programme. This analysis shows that all identified components required under WASH are not exclusively covered by one single framework. Hence, an alternative that can better capture the gender concerns in WASH is needed. This can draw from the strengths of the various frameworks to form a new hybrid framework. For instance, it is very much possible to draw upon the POP, Moser and the Social Relations frameworks to inform issues of participation. This is one of the key indicators to help make the programme gender inclusive. While these frameworks do not help in gauging participation in quantitative figures, they aid in understanding inclusiveness of programmes by looking into different demographic composition as well as social capital aspects (e.g. networks, women's groups) that enhance women's engagement in community activities. The POP and the Capacities and Vulnerabilities Analysis Framework are also useful in drawing attention to various aspects of access to services in terms of hardware access, coverage and the distance covered to avail services.

All frameworks include indicators such as control over land and water sources. However, to understand intra-household management and distribution of water, Moser, GAM and Social Relations frameworks were most helpful. All frameworks, except Capacities and Vulnerabilities Analysis Framework, helped capture benefits in terms of health and livelihoods. The Social Relations Approach best draws attention to institutions at different levels on their influence or governance of gender relations, not captured in the other frameworks. The hybrid framework also needs to capture WASH's O&M aspects in its own right.

Comparatively, the hybrid framework of six indicators makes a WASH programme more gender inclusive and equity based by:

- Assessing inclusion and exclusion parameters to explore underlying linkages between exclusion and gender, with larger impact on WASH.
- Understanding the complexity of gender relations and how this constraints or facilitates addressing gender inequality in WASH.
- Evaluating questions of access to and control over resources, assets and benefits.
- Assessing the barriers and constraints to women and men participating and benefiting equally from the WASH programme.
- Assessing the potential of the WASH programme to empower women and address strategic gender interests to transform gender relations.
- Collecting sex-disaggregated data/information relevant to WASH programme.
- Assessing gender division of labour and patterns on decision-making.
- Monitoring participation, benefits, the effectiveness of gender equality strategies, and changes in gender relations and assessing the gender-sensitive aspects of the WASH programme.
- Building capacity and strengthening networks.

Thus, the hybrid framework can help assess gender differences in participation, benefits and impacts, including progress towards gender equality and changes in gender relations. It provides us with means to assess changes in WASH over time. As it accounts for process and representation, it opens up the scope for active women's participation across diverse communities and social classes. In addition, the framework has an advantage of including community dynamics and networks from the beginning. It makes a clear distinction between two kinds of decision-making processes—within households and institutional set-ups—important to women's empowerment. The framework can help ensure that men and women are not disadvantaged by WASH activities by identifying priority areas for action to promote equality and enhancing sustainability and effectiveness of WASH programmes.

2.7 Locating the Policy and Practice Imperative of the Hybrid Framework

The responses from the Solution Exchange undertaken to support analysis with field experiences revealed a number of successful community-based approaches that target and involve poor and marginalised communities across India. These examples indicate bold moves and paradigm shifts in the way public water supply has been managed, and clean water and sanitation facilities ensured.

Both NGO- and government-implemented programmes have incorporated structured approaches for inclusion of women and the marginalised. These projects and programmes have incorporated gender frameworks to ensure equity and have identified essential phases covering a number of activities. Using the gender framework, corresponding indicators can also be undertaken to ensure equity. These phases can be broadly classified as per Table 2.2.

The Jalanidhi Project, a specific government-partner agency initiative in Kerala, for instance, ensured that 52 per cent of the households belonged to below poverty line (BPL) and 16 per cent from SCs/STs. Over 35 per cent of all key community functionaries were women and under the indigenous community component for STs, 162 small water supply schemes were completed. About 3,700 small schemes with all houses having pipe connections and over 80,000 sanitary latrines were completed. These schemes are completely managed by communities with full O&M cost recovery.

The Jalswarajya Project in Maharashtra, a World Bank sector reform initiative, ensured a policy of entrusting the O&M of rural water supply

Table 2.2:
WASH and gender field-experience collation structure

Phases	*Timeline (Approx.)*	*Broad Activities/Inputs Identified*	*Indicators (Identified by Respondents)*	*Gender Framework Indicators*
Planning and Institutional (Capacity) Building Phase	3–6 months	**Situational Analysis**		
		Situational and gap analysis to assess technical and social feasibility (includes identification of marginalised groups and their needs)	Willingness of the community, their level of participation in awareness campaigns and other activities	Participation (quality of process and quality)
		Community Mobilisation, Involvement and Participation		
		Applying participatory tools (including exposure visits), awareness campaigns to mobilise the community	Attendance of men and women in the *Gram Sabhas*	Participation and inclusion (demographic composition of population)
		Organise *Gram Sabha*/Women's *Gram Sabha* for participatory planning	Formation of village-level committees; number of women represented in local village-level committees, number of people representing vulnerable communities (tribals, Dalits) in the village-level committees	Governance (institutional strengthening in terms of gender, voice)
		Site selection for installation of village hand pumps/paths for piped water supply based on village meetings using the village resource map (the same is applicable if community toilets are to be constructed)		
		Preparation of village action plans		

(Table 2.2 Continued)

(Table 2.2 Continued)

Phases	*Timeline (Approx.)*	*Broad Activities/Inputs Identified*	*Indicators (Identified by Respondents)*	*Gender Framework Indicators*
		Institutional Building		
		Formation of SHGs, village-level WASH committees, and other committees and sub-committees like social audit committee, women empowerment committee, procurement committee, etc.	Number of training programmes/ exposure visits to committee members delineating their specific role and responsibilities; to SHGs for skill development	Benefits to women (livelihood, education)
		Formation, revitalisation and strengthening of community-based organisations (CBOs)	Regularity in committee meetings, records kept on proceedings of the meetings, discussions held and resolutions passed	
		Capacity building/trainings to committee members		
Implementation Phase	9–12 months	**Source Work**		
		Tender process (for initiation of work) and relevant sanctioning of work	Social audit to ensure proper identification and check-up of contractors, materials used, etc.	Participation (quality of process)
		Excavate source or rejuvenate existing source.	Procedure to obtain land, if required	O&M (Participation contribution, monitoring)
		Laying pipelines/hand pumps/water connections	Coverage area, i.e. ensuring it includes all hamlets and vulnerable groups	
		Complete source work.		
		Starting the water supply		

			Number of house connections, hand pumps, etc. provided, number of households with toilet facilities	
		Audit and Handover Final audit and hand over to relevant authority (generally the *Gram Sabha*) for O&M	Number of poor and vulnerable communities excluded from access to water resources (could use the below poverty line criteria)	Inclusion (demographic composition of population) Institutional strengthening in terms of gender
			Hand over to *Gram Sabha* or relevant authority to manage the water supply scheme	
Monitoring and Evaluation Phase	6–9 months	**Operation and Maintenance** Water and electricity charges Water quality and treatment Training of staff involved in O&M Solid waste management	Overdue or balance of tariffs to be paid; Number of households defaulting Prevailing practices of solid waste management (to avoid pollution at source)	Operation and maintenance (participation, contribution and monitoring)
		Monitoring Recovery of water and electricity tariffs Regular water tests Maintenance of source. Regular and proper functioning of all committees formed	Health issues; existing difficulties in accessing water, etc.	Access to services

schemes to women SHGs. This project was started in Khambegaon village, Jalna District, Maharashtra, and was followed by a number of adjoining villages. Presently around 250 villages have engaged women SHGs for daily O&M. The women also collect water tax and are paid a percentage as incentive. A few of the women have also acquired skills in minor repair works.

Covering 173 villages spread over four districts in Gujarat namely, Bhavnagar, Amreli, Panchmahal and Dahod (Utthan), for instance, has been initiating various programmes, including biodiversity for livelihood security in tribal areas, community-led viable alternatives for better management of natural resources, such as sealing of wells, application of organic manure, etc. All these are carried out by various village-level institutions involving women and community-based organisations.

The Jal Bhagirathi Foundation, a local NGO, covers the districts of Jodhpur, Barmer and Pali in Rajasthan. Before initiating water harvesting and natural resource management programmes, sufficient time is dedicated to instil a sense of community ownership. The community is then given the responsibility to search for plausible solutions, detail technical specifications, prepare a supporting budget for the intervention and share them with the members of the foundation. People's enthusiasm and commitment is further consolidated through the *Jal Sabha* [Water User Association (WUA) comprising women members].

Jagori, a local women action group, for instance, has been working in two resettlement colonies in Delhi—Bawana and Bhalswa. Their work has shown that policies and schemes for urban water and sanitation do not cater to women and girls. In these resettlement colonies, women are unable to relieve themselves before nightfall because of fear and shame, which results in high costs of water and sanitation (WATSAN) and related ill-health. Despite high poverty, some families incur debts to build home-based toilets to save girls from harassment and violence.

These examples and associated lessons are compiled in Table 2.2. They show the key phases and associated timelines (a total of 18–27 months is suggested as required) of potential inputs needed from an implementer's viewpoint, for a strong gender equality and equity outcome as well as their indicators to monitor this. This practical and valuable field experience can be linked to the hybrid framework (as per the gender framework indicators; Table 2.2) and the questions for each sub-component referred to earlier can be used by implementers to plan, execute and monitor WASH programmes with this goal. Further research on explicitly linking interventions with identified indicators will help standardise gender evaluation in the WASH sector in future.

2.8 Conclusion

Strengthening gender outcomes in the WASH sector is a critical component to ensure impact and sustainability. In the Indian setting, this becomes even more important, given the central role women play at household levels in domestic water management. It is unfortunate that women are often excluded from WASH interventions despite having an intimate knowledge of the importance of water and how it can positively impact the entire household.

In this chapter, we have offered a review of gender equity in WASH programmes. We have touched upon and identified two gaps currently holding back advances in the WASH-gender discourse in India. The first need is for a detailed gender analysis to better understand how men and women best undertake the planning, decision-making and implementation of WASH programmes. Second, we have reviewed existing gender frameworks to understand how they capture the gender dynamics in WASH. Seven gender development frameworks were selected and examined against six identified areas (participation, access to services, control over resources, benefits to women, governance and O&M and management). Each of these parameters rated the applicability of the various gender frameworks against a series of three to five sub-components.

It was found that all frameworks had gaps in terms of being able to capture all the important issues that the Indian WASH sector comprises. Therefore, a hybrid framework was offered to address these gaps. This was based on a compilation of the selected advantages of the various frameworks. The aim of this hybrid framework is to help strengthen gender-sensitive planning and programming for potential application in the WASH sector in India. The framework suggested comprises a set of core indicators, which can form the basis of developing a corresponding checklist of questions to help baseline development and monitoring in the field.

To further inform the framework review, with practical examples from the field of WASH project implementation in India, a discussion was initiated with grassroots practitioners, via Solution Exchange. The Solution Exchange Water and Gender communities hosted an e-discussion to draw out examples of structured approaches for ensuring inclusion and equity in WASH programmes in India as well as the phases of such approaches. Three key phases of programmes were identified, that is, Planning and Institutional (Capacity) Building Phase, Implementation Phase, and Monitoring and Evaluation Phase. The findings suggest a total timeline of 18–27 months for implementing WASH programmes for improved gender outcomes. A cross-linkage table was developed to show the structure developing from the field examples and the key components of the gender analysis. Further research on explicitly linking interventions with identified indicators will help to standardise gender evaluation in the WASH sector in the future.

Disclaimer

The views expressed herein are those of the authors and do not necessarily reflect the views of UNICEF, the United Nations or the Ministry of Drinking Water and Sanitation, Government of India.

References

Agarwal, A., Narain, S. and Khurana, I. (eds). 2001. *Making Water Everybody's Business: Practice and Policy of Water Harvesting*. New Delhi: Centre for Science and Environment.

Agarwal, B. 1994. *A Field of One's Own: Gender and Land Rights in South Asia*. Cambridge: Cambridge University Press.

Anderson, Mary B., Howarth, Ann M. (Brazeau) and Overholt, Catherine. 1992. A Framework for People-Oriented Planning in *Refugee Situations Taking Account of Women*, Geneva: United Nations High Commission on Refugees (UNHCR).

Anderson, M. and Woodrow, P. 1989. *Rising from the Ashes: Development Strategies in Times of Disaster*. Boulder and San Francisco, and UNESCO, Paris: Westview Press.

AusAid. 2005. Gender guidelines: Water supply and sanitation; Supplement to *The Guide to Gender and Development*.

Bishop-Sambrook, Clare. 2000. *A Manual for Gender-focused Field Diagnostic Studies: IFAD's Gender Strengthening Programme in Eastern and Southern Africa*. Rome: International Fund for Agricultural Development (IFAD).

CARE International Gender Network. 2012. *Good Practices Framework—Gender Analysis*. Virginia: CARE.

Cronin, A.A. and Burgers, L. 2011. Water safety and security: Challenges and opportunities in the Indian perspective, in *Confluence of Ideas and Organisations, International Conference on Water Partnerships towards Meeting the Climate Challenge*, Chennai, India, January 2011, pp. 45–52.

Derbyshire, Helen. 2002. *Gender Manual: A Practical Guide for Development Policy Makers and Practitioners*. London: Department for International Development (DFID).

Goldberg, Toby, Lang, Cathy and C. Lang Consulting. 2004. *Vibrant Communities: Gender Analysis in Community-Based Poverty Reduction: A Report on the Gender and Poverty Project*. Canada: Status of Women Canada.

Jacob, N., Sharma, S., Lala, S. and Mallick, S. 2009. *Consolidated Reply: Impact of School Sanitation on Adolescent Girls—Experiences: Examples*. New Delhi: United Nations Solution Exchange India.

Jiménez, Blanca and Asano, Takashi. 2008. *Water Reuse: An International Survey of Current Practice, Issues and Needs*. United Kingdom: IWA Publishing.

Kabeer, N. 1995. Targeting Women or Transforming Institutions? Policy Lessons from NGOs' Anti-poverty Efforts. *Development in Practice* 5(2): 108–116.

Lala, Sunetra and Basu, Malika. 2012. *Consolidated Reply: Ensuring Inclusion and Equity in WASH Programming—Experiences: Examples*. New Delhi: United Nations Solution Exchange India.

Lerner, Abba Ptachya. 1944. *The Economics of Control: Principles of Welfare Economics*. New York: Macmillan.

Longwe, Sara Hlupekile. 1991. *Gender Awareness: The Missing Element in the Third World Development Project, Changing Perceptions: Writings on Gender and Development*. London: Oxfam.

March, Candida, Smyth, A. Inés and Mukhopadhyay, Maitrayee. 1999. *A Guide to Gender-Analysis Frameworks*. London: Oxfam GB.

MDWS. 2011. *Ministry of Drinking Water and Sanitation, Strategic Plan (2011–22)*. New Delhi: Government of India.

Moser, Caroline. 1993. *Gender Planning and Development: Theory, Practice and Training*. London: Routledge.

Overholt, C., Anderson, M.A., Cloud, K. and Austin, J.E. 1985. *Gender Roles in Development Projects*. Connecticut: Kumarian Press Inc.

Parker, A. Rani. 1993. *Another Point of View: A Manual on Gender Analysis Training for Grassroots Workers*. New York: UNIFEM.

Planning Commission. 2011. *Faster, Sustainable and More Inclusive Growth—An Approach to the Twelfth Five Year Plan* (2012–17). New Delhi.

Rennie, J.K. and Singh, C.N. 1996. *Participatory Research for Sustainable Livelihoods: A Guidebook for Field Projects*. Canada: International Institute for Sustainable Development.

Scanion, John, Cassar, Angela and Nemes, Noemi. 2004. *Water as a Human Right? IUCN Environmental Policy and Law Paper No. 51*. United Kingdom: Thanet Press Ltd.

Seager, Joni, Robinson, Kenza, Schaaf Charlotte van der and Gabizon, Sascha. 2009. *Gender-Disaggregated Data on Water and Sanitation—Expert Group Meeting Report*. Netherlands: Women in Europe for a Common Future WECF.

Status of Women Canada. 2013. *Gender Based Analysis Plus*, http://www.swc-cfc.gc.ca/pol/gba-acs/index-eng.html (accessed 12 April 2013).

United Nations Economic and Social Commission for Asia and the Pacific (UN ESCAP), United Nations Development Programme and Asia Development Bank. 2005. *A Future Within Reach: Reshaping Institutions in a Region of Disparities to Meet the Millennium Development Goals in Asia and the Pacific*. Bangkok: United Nations ESCAP.

United Nations Children's Fund (UNICEF). 2011. *Promoting Gender Equality through UNICEF-Supported Programming in Young Child Survival and Development—Operational Guidance*. New York: UNICEF.

UNICEF, FAO and SaciWATERs. 2013. *Water in India: Situation and Prospects*. New Delhi: UNICEF.

World Health Organisation (WHO) and UNICEF. 2006. *Meeting the MDG Drinking Water and Sanitation Target: The Urban and Rural Challenge of the Decade*. Switzerland: WHO Press.

WHO/UNICEF. 2012. *Progress on Drinking Water and Sanitation: 2012 Update. Joint Monitoring Programme for Water Supply and Sanitation*. New York: UNICEF.

Zwarteveen, Margreet, Ahmed, Sara and Gautam, Suman Rimal. 2012. *Diverting the Flow: Gender Equity and Water in South Asia*. New Delhi: Zubaan.

3

Women and Water: Vulnerability from Water Shortages

Yusuf Kabir, Niranjan Vedantam and M. Dinesh Kumar

3.1 Water Supply Surveillance

Millions of people throughout the developing world use unreliable water supplies of poor quality, which are costly and are distant from their home (WHO and UNICEF, 2000). Over the years, there is a growing realisation that communities in rural areas need water for productive as well as domestic uses, indicating the need for an increase in the quantity and quality of the water supplied from public systems (Renwick, 2008; van Koppen et al., 2006). This is important for meeting United Nation's Millennium Development Goals (van Koppen et al., 2006).

Traditionally, water supply surveillance is used to generate data on safety and adequacy of drinking water supply in order to safeguard human health. Most current models of water supply surveillance come from developed countries and have significant shortcomings if directly applied in a developing country's context. Not only the socio-economic conditions, but also the nature of water supply services is different. Often, water supply services in developing countries comprise a complex mixture of formal and informal services for both the 'served' and 'un-served' (Howard, 2005). For instance, in Maharashtra, only 20.3 per cent of the rural population has access to treated tap water within their dwelling premises. A majority of the rural population depends extensively on private wells, hand pumps, borewells, and ponds and tanks, which provide untreated water, for domestic water supply (Census 2011), a trend found in many other parts of the developing world (Gelinas et al., 1996; Howard et al., 1999; Rahman et al., 1997). The data on actual water use by households and/or communities are absent. However, the sources that are reliable and that can provide adequate quantity of water of good quality to meet various productive and domestic needs seem to be far less than adequate in rural areas, particularly those which are in naturally water-scarce regions. In such regions, the rural poor tend to compromise on their basic needs. Women spend substantial amount of time and effort in collecting water from distant sources. The water is often of poor quality and fails to meet the requirements of drinking and cooking, domestic uses, livestock uses and kitchen garden. In both the cases, there are undesirable outcomes on health and livelihood, particularly on women and children.

Therefore, well-designed and implemented water supply surveillance in relation to domestic and productive needs of the community is important. This should be an integral component to water supply improvement. The key to designing such programmes is information about the adequacy of water supplies and the health and livelihood security risks faced by populations due to lack of it at national or sub-national levels. But, there is a range of natural, physical, social, human, economic, financial and institutional factors influencing the vulnerability of the rural population to problems associated with inadequate supply of water for consumption and production needs (Nicole, 2000). These are not captured in the traditional surveillance programmes.

In this research, we have developed a household-level multiple use water systems (MUWS) vulnerability index in order to assess the problems associated with lack of water for domestic and productive needs for rural communities. This especially focuses on women. We computed the vulnerability index values for 100 sample households each from three villages in Maharashtra.

3.2 Past Approaches to Water Supply Surveillance

The inextricable link between water security, health, livelihood and economic gains is well established (Botkosal, 2009; HDR, 2006). Improving water security of the poor brings about significant health and poverty reduction benefits (HDR, 2006: 42; WHO, 2002). The economic losses due to deficit in water supply of sufficient quantity, quality and reliability are disproportionately higher for poor communities. This is owing to greater risk of employment loss, health costs, loss of productive workforce and water-based livelihoods (HDR, 2006: 42).

As Nicole (2000) argues, a demand-responsive approach to water supply requires that the livelihood needs of the community are also taken into account, rather than the supply requirements for human consumption and sanitation needs. Therefore, an assessment of water supply at the household level worked out on old norms surrounding the notion of water supplies that meet human health and hygiene needs is grossly inappropriate. In India, the monitoring of rural water supply is based on simplistic considerations, involving data on number of households covered by different types of water supply systems and the characteristics of the sources. The data gathered through such surveys are silent on the amount of water *actually* consumed by the population, and the quality and reliability of the supplied water, all of which determine the health and livelihood outcomes.

3.3 Why Vulnerability Index for MUWS?

The foregoing discussion suggests that comprehensive approaches to water supply surveillance were by and large lacking for quite some time. The approaches to water supply surveillance that allow targeting of surveillance activities on vulnerable groups were assessed by G. Howard using case studies from Peru and Uganda. The Peru case study attempted to incorporate some measures of vulnerability into the surveillance programme design through a process of 'zoning' that was based on water service characteristics. The Uganda case study involved the development of a semi-quantitative measure of community vulnerability to water-related diseases, to zone urban areas and plan surveillance activities. The zoning used a categorisation matrix, which was developed incorporating a quantitative measure of socio-economic status, population density and a composite measure of water availability and use (Howard, 2005).

But, the main limitation of the Howard approach is that it assessed the vulnerability households against lack of water for human consumption and sanitation. It did not take into account the multiple water needs of the community, particularly of the poor in rural areas. There are many factors such as the family occupations, social profile and financial stability, which determine the household water needs for productive purposes that were not accounted for.

Identifying the most vulnerable groups is not an easy task due to the complex interplay of a wide range of factors. Factors such as poor reliability (continuity of supply), costs (affordability) and distance to the source may all lead households to depend on less safe sources. It reduces the volume of water used for hygiene purposes and spending on other essential goods, such as food (Cairncross and Kinnear, 1992; Howard, 2002; Lloyd and Bartram, 1991). The evidence suggests that water interventions targeted at poor populations provide significant health benefits and contribute to poverty alleviation (DFID, 2001; HDR, 2006; WHO, 2002). Though it appears that poverty is a major factor deciding vulnerability, it is just one of the many complex factors that eventually determines the outcomes of family's high vulnerability to lack of water for multiple uses.

The factors that can influence vulnerability of a household to problems associated with lack of water for multiple uses could be: (1) degree of access to water supplies for human consumption, personal hygiene, and productive uses such as livestock consumption in terms of quantity and desired quality, and the level of use; (2) social profile and family occupations; (3) social institutions and ingenuity; (4) condition of water resources; (5) climatic factors; and (6) financial condition (Lloyd and Bartram, 1991; Cairncross and Kinnear, 1992; DFID, 2001; Howard, 2002; Hunter, 2003; Nicole, 2000; Sullivan, 2002; WHO, 2002). Poor access to water supply sources directly impacts women, as it increases their effort to fetch water. Poor quality of water and low level of use increase water-related health risks of children. The second and fifth factors influence the vulnerability by changing the household water demand. This may not be always in terms of the quantum of water, but in terms of the reliability of supply. The third and fourth factors can change the external environment, which influences water supply. Here again, the degree of access depends on the presence/absence of social institutions and local customs and traditions prevalent in developing countries.

Climate has a major bearing on the adverse effect of lack of water for hygiene and environmental sanitation. In arid and semi-arid climates, breeding of water-related insect vectors is less during hot weather conditions. In flood-prone areas and areas receiving high rainfall, the occurrence of water-based diseases are likely to be more, and therefore more caution needs to be exercised in the disposal of human and animal excreta (Hunter, 2003: 37). At the same time, the demand of water for meeting livestock needs, and irrigating fruit trees and kitchen garden etc., would increase with increase in aridity and temperature. So is the demand for water for washing and bathing. Arid areas are also drought-prone. Hence, there is a need to develop a composite index that takes into account these complex factors in assessing the vulnerability of rural households to inadequate supplies of water to meet multiple needs so as to make surveillance more targeted.

3.4 Objectives and Methodology

Rural water supply schemes in India are generally designed for single use, that is, domestic use. The multiple water use priorities of poor rural households, especially women in order to reduce their hardship, and enhance food production, health and income means that in water-scarce areas, the domestic water use can run into conflict with productive water use. The goal of this research study based in Maharashtra is to develop replicable models of MUWS. This will provide year-round access to water for domestic and productive uses under varying climates for vulnerable households. The site comprises three *Gram Panchayats* (village councils) of rural Maharashtra with a special focus on disadvantaged women. The research study started with the following intentions: identification of various domestic and productive water needs of village households; development of a composite index capturing the vulnerability of rural households to problems associated with lack of water for multiple needs; assessment of multiple use vulnerability of sample rural households in the three villages of Maharashtra; analysis of socio-economic and livelihood dynamic of the sample households, with particular reference to the impact of climate variability on the household. Documentation of the key steps followed for designing of MUWS models are suited to the three different settings, and the description of the final outcomes; and, working out the institutional arrangements for implementing these MUWS systems.

3.5 Deriving an MUWS Vulnerability Index

The vulnerability of a household to inadequate supply of water to meet drinking water, sanitation and livelihoods needs is determined by four broad parameters: (1) capital assets and good; (2) sequencing and time; (3) institutional linkage and (4) knowledge environment. The capital assets can be further divided into natural capital, social capital, physical capital and financial capital (Nicole, 2000). It is evident that while some of the capital assets (physical capital-related and human capital-related parameters) would determine the access to water supply and its use, the natural capital-related parameters, institutional linkage and knowledge environment would change the external environment that influences the supply and use for water. The capital assets such as natural capital, social capital and financial capital influence the demand for water.

All these parameters are factored in six broad sub-indices, we have discussed in the previous section. Each one of these six broad factors constitutes one sub-index. The number of 'minor' factors, which together are considered to be influencing the measure of these sub-indices, the methods and procedure for their computation, and sources of data are explained in Table 3.1. The composite index of 'MUWS vulnerability' will have a maximum value of 10.0, meaning zero vulnerability; lower values of the index mean higher vulnerability. It is composed of six sub-indices (from A to F in Table 3.1). Each one has unequal weightage in deciding the value of the index. The maximum value of sub-index A is 3.0; that of B, C and D will be 1.0; and that of E and F will be 2.0 each. The sub-sub index has equal weightage (measured on a scale of 0–1.0). The sum of values of all sub-indices under sub-index A is multiplied by 0.3. This is to obtain the value to be imputed into the mathematical formulation for estimating the composite index. The sum of the values of all sub-indices under sub-index 'B' and 'D' is divided by 2 to obtain the value to be imputed into the mathematical formulation for estimating the composite index. The sum of sub-sub indices under sub-index 'E' is multiplied by

Table 3.1:
Deriving a household-level MUWS vulnerability index

S. No.	Parameters	Quantitative Criterion for Measurement	Method of Data Collection
A	**Water supply and use**		
1	Access to water supply source (primary)	Vulnerability decreases with improved access. Access is an inverse function of the distance. The index is a function of the distance to the source from '0' within the dwelling to a maximum of 1 km and above in gradations of 0.2[1]	Primary survey
2	Frequency of water supplies	Vulnerability increases with decrease in frequency of water delivery[2]	Do
3	Ownership of alternative water sources	Ownership of an alternative water source would increase the overall access and reduce the vulnerability[3]	Do
4	Distance to the alternative source 'owned'	Distance to the alternative source would increase the vulnerability. Often, the alternative sources are farm wells, which are located outside the village[4]	Do
5	Access to other alternative sources	Vulnerability decreases with number of alternative sources[5]	Do
6	Capacity of domestic storage systems	Vulnerability to lack of regular water supplies decreases with increase in volume of storage systems in place[6]	Do
7	Quantity of water used	The vulnerability increases with decrease in quantum of water used against the requirement. The vulnerability can be treated as zero when all the requirements in the household are fully met[7]	Do
8	Quality (chemical, physical and bacteriological) of domestic water supplies	Poor quality of drinking water increases vulnerability; Bacteriologically, physically and chemically pure is the best water[8]	Lab test results/ perceptions
9	Total monthly water bill as a percentage of monthly income	Vulnerability increases with increasing percentage of total family income spent on water. An expenditure level of 10 per cent of monthly income is treated as highest and most vulnerable[9]	Primary survey

(Table 3.1 Continued)

(Table 3.1 Continued)

S. No.	Parameters	Quantitative Criterion for Measurement	Method of Data Collection
B	**Family occupation and social profile**		
1	Family occupation	Vulnerability will be low for families having regular source of livelihood that are not dependent on water. Those who are dependent on irrigated crop production are considered to be not vulnerable. But, those who are dependent on dairying will be vulnerable. The vulnerability will reduce if they depend on wage labour and other sources of livelihood that do not require water[10]	Do
2	Social profile	Vulnerability is also a function of the social profile. The families having school-going children are more vulnerable to inadequate quantity, quality and reliability of water supplies. So, is the case with families having office-going adult. But, the vulnerability would reduce with the presence of surplus labour availability[11]	Do
C	**Social institutions and ingenuity**	Community's vulnerability to the problems associated with lack of water increases in the absence of social/community institutions; social ingenuity[12]	Primary survey (but qualitative to be obtained from discussions)
D	**Climate and drought proneness**		
1	Climate (whether semi-arid/arid/hyper-arid or sub-humid/humid)	The vulnerability to lack of water for environmental sanitation is a function of climate. It increases from hot and arid to hot and semi-arid to hot and sub-humid to hot and humid to cold and humid[13]	Secondary data on climate
2	Aridity and drought proneness	The vulnerability due to lack of water for domestic uses, livestock increases with increase in aridity as it would increase the demand for water for washing, bathing, livestock drinking and irrigation of vegetables and fruit trees. Aridity areas are also drought-prone[14]	Do

(Table 3.1 Continued)

(Table 3.1 Continued)

S. No.	Parameters	Quantitative Criterion for Measurement	Method of Data Collection
E	**Condition of water resources**		
1	Surface and groundwater availability in the area	A renewable water availability of 1,700 m^3 per capita per annum is considered adequate for a region or town, estimated at the level of river basin in which it is falling.[15]	Secondary data
2	Variability in resource condition	Higher the variability, greater will be the vulnerability[16]	Do
3	Seasonal variation	Regions that experience high seasonal variation in water availability are highly vulnerable[17]	Do
4	Vulnerability of the resource to pollution or contamination	Surface water is more vulnerable to pollution than groundwater. Shallow aquifer is more vulnerable than deep confined aquifer[18]	Do
F	**Financial Stability**	Overall financial stability of the family would influence the vulnerability. This is different from the earnings from current occupations. The savings in banks/post office; ownership of productive land, which is not mortgaged[19]	Primary survey

0.5. Here more weightage is given to 'A', as it contains nine sub-indices. This is not the case with other indices 'B', 'C', 'D' and 'E'.

3.6 Computation of MUWS Vulnerability Index for Three Villages in Maharashtra

A preliminary survey was carried out in six villages from three locations in Maharashtra. The three locations for project were selected in such a way that they represent three distinct typologies. They are as follows: (1) a village located at the foot of the hilly region, characterised by high rainfall and plenty of local streams flowing down from high altitudes fed by base flows from hilly aquifers; (2) a village located in hard rock plateau areas with low-to-medium rainfall, with the rural water supplies heavily dependent on the limited groundwater resources in the Deccan trap formations; and (3) a village located in the foot of hilly forested land, falling in the assured rainfall zone, with extremely limited groundwater, but has local streams.

From these six villages, three were selected for the action research. Following physical and socio-economic criteria were used in the selection process: (1) maximum number of households in the village have access to domestic water supply through tap connections (where the physical access to water for drinking and cooking and other household uses is very good); (2) predominantly agrarian villages where a significant section of the farm households are not able to meet their farming needs from the available water sources (ponds, tanks and wells), and therefore demand water for multiple purposes, including water for livestock, water for kitchen garden to improve their livelihoods, from the public systems; and (3) the current public water supplies across seasons are less than adequate to meet these needs largely due to competition from agriculture, but conditions favourable for augmenting the available supplies, through technological and institutional measures.

The selected villages are: Varoshi (Jawali taluka of Satara district), Kerkatta (Latur taluka of Latur district) and Chikhali (Jivati taluka of Chandrapur district). They represent three different agro-ecological zones in the state, one from the high-rainfall zone of western Maharashtra located in the foothills of Western Ghats, with high humidity; another from the low-to-medium rainfall zone of central plateau, which is drought-prone and experiencing high aridity; and the third from the assured rainfall zone of Vidarbha, at the foot of hills, with rainfall exceeding 1,400 mm. All the villages face problems of inadequate availability of water for meeting multiple needs such as animal drinking, vegetable cultivation throughout the year and water for basic needs during summer months. Quality and reliability of water are not issues in these villages.

Varoshi falls in the high rainfall region, Kerkatta in the low rainfall, drought-prone region, and Chikhali in the assured rainfall region. The MUWS vulnerability index was computed for 100 sample rural households each in three villages of Maharashtra. The household level data on the range of physical and socio-economic variable affecting the MUWS vulnerability were collected through a survey.

3.7 Results

In the case of Varoshi, the value of the index was found to be varying from 3.31 to 6.58. Figure 3.1 shows the values for all the sample households. Out of the 100 households surveyed, 67 households have vulnerability index values lower than 5, which means these are more vulnerable. Varoshi, despite being in a high-rainfall region, had more households vulnerable. This can be attributed to the fact that, the main source of water is a spring needing repairs. The villagers were facing water scarcity during two months of summer.

In the case of Kerkatta village as shown in Figure 3.2, the computed values of multiple use vulnerability index at the household level for the sample households range from a lowest of 2.21 to a highest of 6.32. Out of the 100 households, 81 are having values lower than 5.0, and therefore are considered vulnerable from multiple water use point of view. As Kerkatta village falls under low rainfall and drought-prone area in Maharashtra, our results show more households are vulnerable.

Figure 3.1:
Multiple use vulnerability index for sample households, Varoshi, Satara

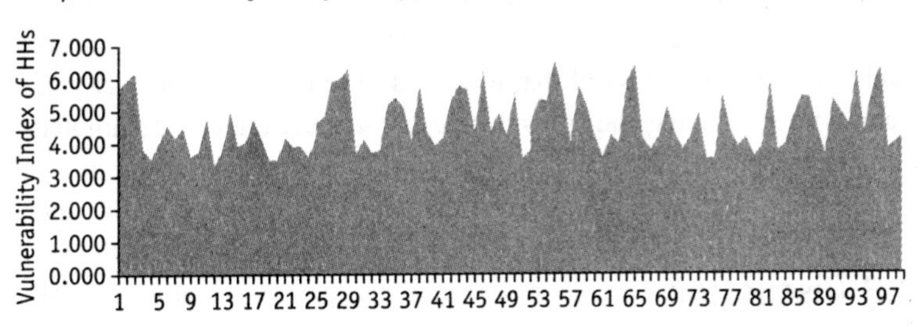

Figure 3.2:
Multiple use vulnerability index for sample households, Kerkatta, Latur

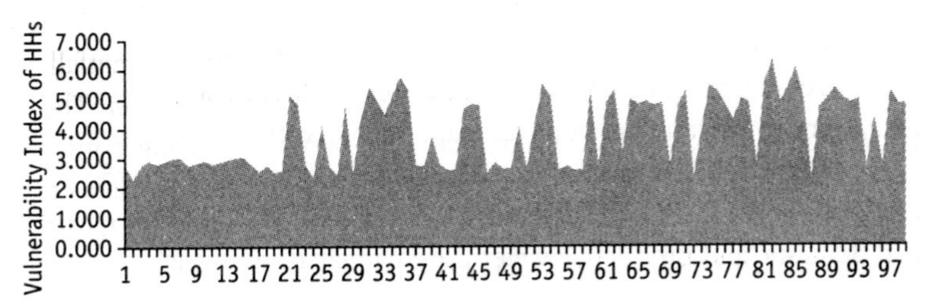

Figure 3.3:
Multiple use vulnerability index for sample households, Chikkali, Chandrapur

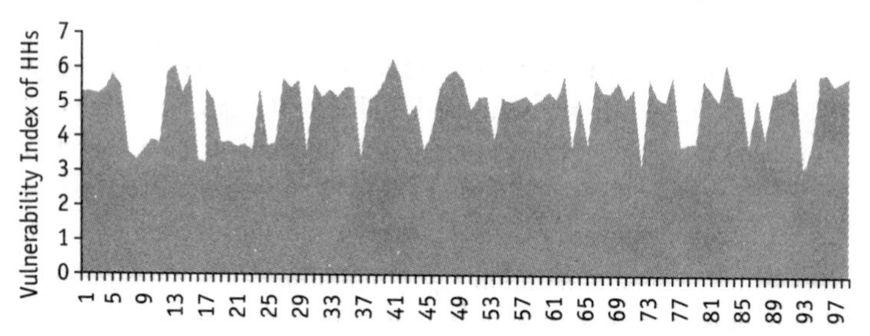

On the other hand, in Chikkali village, which is an assured rainfall area, the computed values of multiple use vulnerability index at the household level (see Figure 3.3) for the sampled households range from a lowest of 3.15 to the highest of 6.37. Out of the 100 households, 30 have vulnerability index lower than 5.0, and hence are treated as highly vulnerable.

Hence, out of these three villages, Kerkatta had the maximum percentage of households vulnerable to problems associated with inadequate water supply for multiple uses. Women and adolescent girls have been observed to spend more time in fetching water during the lean seasons. The average distance travelled to collect water during rainy season, winter season and summer season are 254.4, 254.4 and 566.4 metres, respectively. Also, it was observed that women and school-going girls sacrifice water for personal hygiene needs during summer seasons to accommodate other domestic needs.

3.8 Discussion

The proposed MUWS vulnerability index is a composite index that takes into account several complex factors such as social, physical, human and economic in assessing the vulnerability of rural households to inadequate supplies of water to meet multiple needs. This can help in designing water management programmes and can make surveillance more targeted and effective. The analysis of MUWS vulnerability index from using the primary data in the three villages showed that households in all three villages had domestic and productive water uses, namely drinking and cooking;

bathing, washing and toilet use; providing drinking water for livestock; and watering of kitchen gardens. The household vulnerability index computed post-intervention showed some reduction in the vulnerability of the households to problems associated with lack of water for multiple needs in terms of reducing the time of fetching water for the women and adolescent girls and improved sanitation and hygiene practices. The highest reduction was found in the case of Kerkatta village, which had the highest number of vulnerable households prior to the intervention. The women in Kerkatta village were observed to be having more control with daily water budgeting and using water for kitchen gardening, poultry, etc. Regular bathing in lean period was not seen as a luxury by some of the adolescent girls in the habitation. This project was also able to increase the hygiene and cleanliness awareness during menstruation, as very often it was observed that dirty water was being used for cleaning of cloths.

3.9 Conclusion

This chapter covered the key outcomes of a pilot project undertaken on developing and implementing an MUWS in sample villages of Maharashtra. The intention was to address multiple needs of the community with a focus on women's management of scarce water resources at household level. The research phase consisted of the following: (1) a detailed survey of sample households in the villages to assess the different domestic and productive water needs of the households, to analyse various dimensions of water to and use of water, and to assess their socio-economic and livelihood dynamics, with particular reference to women and the impact of climate variability on the same; (2) to assess the multiple use vulnerability of the households; (3) an extensive review of various multiple use models available from around the world with particular reference to the study of features that contribute to their good performance; (4) design MUWS that suit each one of the three agro-climatic and socio-economic settings; (5) an extensive review of effective water institutions from around India; and (6) developing institutional arrangements for managing MUWS in the pilot villages.

Developing a multiple use vulnerability index at the household level was one of the exercises carried out under the research phase. Six broad sub-indices constitute the composite multiple use vulnerability index developed by us. They are: (1) water supply and use; (2) family occupation and social profile; (3) presence of social institutions and ingenuity; (4) water resource endowment; (5) climate and drought proneness; and (6) financial stability. These sub-indices broadly capture the key parameters like capital assets (natural, physical, social, financial), sequencing and time, institutional linkages, and knowledge environment. This index helped assessing water issues of vulnerable rural population and was useful in water supply surveillance. In this research, we have developed a household-level MUWS vulnerability index to assess the problems associated with lack of water for domestic and productive needs for rural communities, and subsequently computed its values for 100 sample households each from three villages in Maharashtra. For this, primary data were collected through a household survey. Among the three villages, Kerkatta in Latur district had the highest number of vulnerable households in the sample. Identification of vulnerable households helped us suggest improvements in water supply to reduce the hardships faced by women in fetching water from distant and unreliable sources. The results emerging from computing the vulnerability index are in line with the intentions of the study. The computed vulnerability index was used for design of water supply augmentation schemes and retrofitting of existing water supply infrastructure to meet the multiple water needs of poor rural households with special focus on women.

Disclaimer

The views expressed herein are those of the authors and do not necessarily reflect the views of UNICEF or the United Nations.

Notes

1. Within the dwelling is '1.0'; within the premise is '0.80'; within 0.2 km distance is '0.60'; between 0.2 and 0.5 km is '0.4'; 0.5 and 1.0 is '0.2' and more than 1.0 km is '0'.

2. Frequency can be indexed as total hours of water supply in a week as a fraction of number of hours.
3. It is assumed that the ability to manage water is the highest in the case of a functional open well, followed by borewell, hand pump and farm pond in the decreasing order. The value of the sub-index would be 1 in the case of ownership of a functional open well, followed by 0.70 for ownership of a borewell; 0.50 for ownership of a hand pump; and 0.3 for ownership of a farm pond.
4. Within the premise is '1.0'; within 0.2 km distance is '0.80'; between 0.2 and 0.5 km is '0.6'; 0.5 and 1.0 is '0.4' and more than 1.0 km is '0.20'.
5. The value of sub-index for this attribute would be '1.0' if there are four alternate sources and above, and the value would decrease proportionately with decrease in number of alternative sources.
6. It would decrease with increase in the ratio of the 'actual storage capacity available' to the 'storage capacity required'; and the value of the index would be higher. The storage capacity required would be an inverse function of the frequency of water supply. If supply comes once daily but during odd hours, then it can be assumed that the volume of water for the entire day's use would be required to be stored. So, the storage capacity would be '$n \times f$'. If it comes during daytime for less than an hour, then half the daily water use would be the storage requirement. For more than one hour, the storage requirement would be minimal (around 20 litres per capita). With alternate day water supply, it could be $2 \times n \times f$. For once in three days, it would be $3 \times n \times f$ and likewise. For round-the-clock water supply, the storage requirement would be zero, and here the ratio can be assumed as 1.
7. This sub-index is computed by taking the volume of water used (x) as a fraction of the minimum required (n), i.e. $\frac{x}{n}$, where n is water requirement as per norms. The value of n should be estimated by considering the human requirement of 50 lpcd (basic survival need as suggested by Glieck, 1997); the animal requirements decided by the types of livestock and the size; and the requirement for kitchen garden.
8. The value of the sub-index 'm' would be 0.33 if the water is pure either bacteriologically or physically or chemically. The value would be 1.0 if pure on all counts.
9. The value of the sub-index would be '0' if the family spends 10 per cent or more of its monthly income on obtaining domestic water supplies, and would keep on increasing with reducing amount of money spent in water bill. The mathematical formulation for computing the index therefore is $[1 - W_C/MI]$, where W_C is the monthly expenditure on securing water supplies and MI is the monthly family income.
10. The vulnerability induced by family occupation is considered to be zero, if the adults in the family are engaged in livelihoods that are not dependent on water in the village. The vulnerability is also considered to be zero for families

having crop production with own irrigation facilities. The families purely dependent on dairy farming would be assumed to have highest vulnerability (1.0), as the water for cattle drinking will have to be managed.

11. For families having school-going children and office-going adults, the situation could be treated as most unfavourable. Here, the sub-index could be assumed as 0.0 (lowest), meaning highest vulnerability. The families having either of these, the value could be assumed as 0.33. For families having neither of these, the value would be treated as 0.67. For families, having surplus labour in the HH for fetching water from distance, the sub-index could be assumed as 1.0.
12. The value can range from '0' for the absence of social institutions or ingenuity to 0.5 for presence of either of these to 1.0 for the presence of both. Social institutions would include: WATSAN committees (Y = 0.50; No = 0.0). Social ingenuity would include existence of water sharing traditions between households during crisis (Y = 0.25; No = 0.0) and practice of re-using water in households using bathing/washing water for toilet flushing, use of sand and ash for cleaning utensils, etc. (Y = 0.25; No = 0.0).
13. The value ranges from '0.0' for cold and humid to '1.0' for hot and arid with increments of '0.20'.
14. The value ranges from '1.0' for cold and humid to '0.0' for hot and arid with reduction of '0.20'.
15. A renewable water availability of 1,700 m^3 per capita per annum is considered adequate for a region or town, estimated at the level of river basin in which it is falling. The value of the index is computed by taking the amount of renewable water as a fraction of the desirable level of 1,700 m^3.
16. The index is computed an inverse function of the coefficient of variation in the rainfall variability in that basin/sub-basin ($1 - x/100$), where x is the coefficient of variation in rainfall.
17. For alluvial areas, the value of this index is considered as 1. For hard rocks, the value is considered as 0.3. For sedimentary and alluvial deposits, the value is considered as 0.65.
18. Shallow groundwater areas and river/stream/reservoirs in the vicinity of industries are highly vulnerable with a value of the sub-index equal to 0.0; distant reservoir in the remote virgin catchments and groundwater from deep-confined aquifers have a pollution vulnerability index of 1.0; shallow groundwater in rural areas have medium vulnerability with a value of 0.50.
19. The family having 1.0 ha of productive land per member in a semi-arid, water-scarce region or 0.5 ha of productive land per member in a water-rich area is considered to be financially stable with zero vulnerability; the vulnerability is assumed to increase gradually with reducing size of land owned, the highest vulnerability being for landless. Again, income savings can compensate the lack of ownership of land, with a total saving of ₹20,000 equivalent to 0.5 ha in water-rich area and 1.0 ha in a water-scarce area.

This index could be computed as 'x/0.50' for water-rich areas and 'x/1.0' for water-scarce areas (where x is the land owned in hectare).

References

Botkosal, W. 2009. Water Resources for Livelihood and Economic Development in Cambodia, Centre for River Basin Organizations and Management, Solo, Central Java, Indonesia.

Cairncross, S. and Kinnear, J. 1992. Elasticity of demand for water in Khartoum, Sudan. *Social Science Medicine*, 34(2): 183–189.

DFID. 2001. Addressing the Water Crisis: Healthier and More Productive Lives for Poor People, Department for International Development, London.

Gelinas, Y., Randall, H., Robidoux, L. and Schmit, J.P. 1996. Well water survey in two districts of Conakry (republic of Guinea) and comparison with the piped city water. *Water Research*, 30(9): 2017–2026.

Glieck, P.H. 1997. Human population and water: Meeting basic needs in the 21st century, in Pachauri, R.K. and Qureshi, L.F. (eds), *Population, Environment and Development*. New Delhi: Tata Energy Research Institute, pp. 105–121.

Howard, G. (ed). 2002. *Water Supply Surveillance: A Reference Manual*. Loughborough, UK: WEDC.

Howard, G. 2005. Effective water supply surveillance in urban areas of developing countries. *Journal of Water and Health*, 3(1): 31–43.

Howard, G., Bartram, J.K. and Luyima, P.G. 1999. Small water supplies in urban areas of developing countries, in Cotruvo, J.A., Craun, G.F. and Hearne, N. (eds), *Providing Safe Drinking Water in Small Systems: Technology, Operations and Economics*. Washington, D.C.: Lewis Publishers, pp. 83–93.

Human Development Report (HDR). 2006. Human Development Report—2006, New York: United Nations.

Hunter, P.R. 2003. Climate change and water-borne and vector borne diseases. *Journal of Applied Microbiology*, 94: 37–46.

Lloyd, B. and Bartram, J. 1991. Surveillance solutions to microbiological problems in water quality control in developing countries. *Water Science and Technology*, 24(2): 61–75.

Nicole, A. 2000. Adopting a Sustainable Livelihoods Approach to Water Projects: Implications for Policy and Practice, *Working Paper 133*, Overseas Development Institute, London.

Rahman, A., Lee, H.K. and Khan, M.A. 1997. Domestic water contamination in rapidly growing megacities of Asia: Case of Karachi, Pakistan. *Environmental Monitoring and Assessment*, 44(1–3): 339–360.

Renwick, M. 2008. Multiple Use Water Services, Winrock International, GRUBS Planning Workshop, Nairobi, Kenya.

Sullivan, C. 2002. Calculating Water Poverty Index. *World Development*, 30(7): 1195–1211.
Van Koppen, B., Moriarty, P. and Boelee, E. 2006. Multiple Use Water Services to Advance the Millennium Development Goals, IWMI Research Report 98. IWMI: Colombo.
World Health Organization. 2002. *Reducing Risks, Promoting Healthy Life, World Health Report 2002*. World Health Organization, Geneva.
WHO and UNICEF. 2000. Global Water Supply and Sanitation Assessment (2000), Report. World Health Organization/United Nations Children's Fund, Geneva/New York.

4

Crossing Boundaries: Gender and IWRM in Education and Research

Anjal Prakash and Chanda Gurung Goodrich

4.1 Introduction

While new research and innovation in various water disciplines are important, the notion of 'integration' remains elusive in water-related projects in South Asia. This is particularly so from the natural/technical and social scientific perspectives. Innovative research and training is therefore needed to enhance the knowledge base of gender-inclusive integrated water resources management (IWRM). Such knowledge is best developed in the context of real water resources management problems, and efforts at intervention, transformation and reform towards achieving integrated resource management. This chapter documents the experiences of a project—The Crossing Boundaries (CB)—that was implemented in South Asia during 2005–2011. The project sought to build capacities of water professionals in IWRM through higher education, innovation and social learning-focused research for gender and water in South Asia. It was implemented by six partner institutions in Bangladesh, India, Nepal and Sri Lanka, and was coordinated by the South Asian Consortium for Interdisciplinary Water Resources Studies (SaciWATERs),[1] based in Hyderabad, India, and the Irrigation and Water Engineering Group at Wageningen University, the Netherlands.

Based on the ideology of both north-south and south-south collaboration, this project focused on education, research/innovation, knowledge base development and networking in a combined effort to contribute to a paradigm shift in water resources management in South Asia. This focus on longer duration education input, as opposed to short-term training, derived from the fact that shaping attitudes and perceptions, and teaching the skills of interdisciplinary, comprehensive analysis and intervention requires adequate time. The project was implemented by a group of institutions with a proven interest and track record in integrated, interdisciplinary and gender-sensitive approaches to water resources management. The CB initiative was built on regional cooperation between researchers across South Asia with common professional interests on the issue of IWRM and gender and water. To ensure that research activities were relevant, they were explicitly linked or aligned to ongoing development projects such as National Employment Guarantee Scheme in India and Community-Based Total Sanitation Efforts in Bangladesh. The experience shows that even with limited funds, large impacts can be achieved and the issue of gender can be addressed through planned capacity-building programme in the South Asian context.

This chapter documents this process and outlines the challenges faced in implementing an interdisciplinary approach to water education and research in South Asia. This chapter is divided into three parts. Part one provides the context of South Asia's water education. Part two lays out the project initiatives to address the problems. The final part discusses the outputs and outcomes of the initiative and outlines major challenges.

4.2 The Context

Women's status and their access to employment and education are poor in South Asia. Women's involvement in water-related education and their participation in water-related workforce are low. But women's mostly unremunerated contribution to the water sector is high; so is the potential impact of enhanced access to water for women's livelihoods (Ahmed, 2005). Gender and water issues encompass not only the roles and activities of women in the public, academic and private sector but also that of women activists working in community mobilisation in water-related issues (Krishnaraj, 2011). Recent international focus on gender relations in the water sector has led to gender mainstreaming in the water sector, mostly through donor-funded projects (Joshi, 2005, 2011). But in some cases, gender mainstreaming has unfortunately been limited to cosmetic interventions supporting elite groups and has not resulted in any positive enhancement of equity or improved women's access to decision-making (Joshi and Zwarteveen, 2011; Kapadia, 2002). In other instances, gender-blind projects can actually increase women workload without enhancing benefits (Lahiri-Dutt, 2006; Zwarteveen et al., 2012).

So where do the bottlenecks lie? Apart from the diverse set of issues determining lack of access of women for water services, a major and understudied area is to recognise the nature and role of water professionals and their educational training (Cap-Net, SOPPECOM, WWN, 2007). Most of the water professionals come from civil engineering background, which largely focuses on technical issues surrounding water and sanitation. Gender and equity issues in water management hardly form a part of their curriculum; sanitation may have a small component and hygiene may not feature at all. Therefore, part of the problem lies in the orientation of the water professionals with their educational background being techno-centric, gender-blind with very limited field orientation.

SaciWATERs initiated studies to understand the course curriculum of leading engineering education institutions in four South Asian countries in the years 2000 and 2008 to understand the changes in curriculum and training of engineering students (Mollinga, 2009). The organisation found that since 2000, water resources education programmes in the South Asian subcontinent has become diverse. In all the four countries studied, Bangladesh, North India and South India, Nepal and Sri Lanka, water resources engineering and hydrology-focussed education dominate. The study found that there certainly is no discernible massive change in the orientation of higher water resources education. The initiatives do cater to address the challenges of water sector—ecological sustainability, securing livelihoods, and eradication of poverty and democratic governance. Among these, ecological challenge is receiving the most response. There is a clear observable trend of stronger focus on the environmental and ecological aspects of water resources management. This partly happens within existing engineering and hydrology programmes, but more prominently in environmental studies programmes that also focus on water resources. In the past decade, a substantial number of new environment-oriented study programmes with a focus on water resources have been established in Bangladesh. Similarly, in Sri Lanka, the number of degree programmes focusing on environmental engineering and management has clearly increased over the past decade. The second and third challenge, livelihood security/poverty alleviation and democratic governance, come some distance behind that of ecology/environment. They are still marginally addressed. The striking feature is that there are no real signs that 'gender and water' is receiving more systematic attention in water resources education, except for some isolated courses (Mollinga, 2009).

So, techno-centric education is still the dominant paradigm for water resources teaching and education in South Asia.

The ingredient of this techno-centric education entails gender blindness in the name of gender neutrality. Most of the courses offered do not have any analysis of gender and equity issues, which stems from the fact that water is treated as a masculine subject. Not only are the courses gender-blind, but the infrastructure that these engineering colleges have is a predominance of male faculties and male students (Kulkarni, 2012). There are very few courses that engage with ground realities and local contexts. The fact remains that there is still a long way to go and many boundaries to cross to achieve a critical mass of education programmes that incorporate the new water resources management challenges. Though one cannot expect all water resources education programmes to become interdisciplinary and integrated at once, there is a need for specialised programmes. This need applies to the natural and engineering disciplines, as well as in the environmental, social, economic, political and geographical disciplines. However, for 'integration' to become real and have a broader impact on policy and in society, a larger number of institutes and programmes needs to address the integration question explicitly, in teaching as well as research, as research informs curricular content.

4.3 The Crossing Boundary Project Initiative: How Were the Issues Addressed?

The CB project has addressed these issues in a multi-pronged approach, targeting the curricula, capacity building of human resources (faculty and students), influencing research through participatory research methodology as well as taking up outreach and advocacy activities. We explain each of these initiatives here.

4.3.1 Curricula

To address the techno-centric, non-interdisciplinary and classroom-oriented approach of the education system, the CB project had taken two initiatives. Firstly, it developed and introduced interdisciplinary courses.

Secondly, it developed new modules. The project developed and introduced three new courses—IWRM, Gender and Water, and Field Research Methodology. The courses were developed jointly by the faculties of the institutions involved and the project staff and experts in these fields who were hired as resource persons. The first step was conducting a training workshop for the faculties on the subjects where the concepts, principles and approaches were presented, discussed, debated and key themes pertaining to South Asia identified. This was followed by field visits and observations, based on which reflections, discussions and debates were held. Finally, the faculties, with assistance from the experts in the subjects and developed modules and courses tailored to the requirements of respective institutes. These were then presented to all the participants for their feedback. The three courses were included in the curricula of the four universities that were part of CB initiative. In addition to the new modules developed and incorporated in student's curricula, staff trainings were conducted on newer issues relating to the three main subjects. The topics covered a wide range of issues such as water and equity, water and heath, water and economics, participatory field research methodology, water and ecosystems, and water and climate change.

4.3.2 Capacity Building of Faculty and Students

In the area of human resource development, the project had two objectives. Firstly, to build a cadre of women water professionals (WWPs) who understand and appreciate the complexity of the concept of IWRM and the challenges involved in its practice. Further, they would have the skills for interdisciplinary approach and participatory field research. Secondly, the aim was to build and strengthen the capacity of the faculties of these institutions in the subjects and new issues. The strategies adopted to achieve these objectives were through provision of special fellowships awarded to girl students for pursuing master's and PhD courses. The other aspect of capacity-building initiative was to train the existing faculties. The project conducted two staff training per year on different aspects of water. The training design was suited to adult learning mode and was interactive and experimental. Similarly, the students supported by the scholarship were trained at South Asia level through a two-week module that focused on participatory methods for research and action. The objectives of the training were to develop an understanding of the concept and principles of participatory research, to practise methodologies/tools for participatory field research and for learning the importance of effective participatory field research. The programme covered concepts, methods and tools for participatory field research, including field practice that enhanced the research capability of engineers for interdisciplinary water resource management in the region.

In the first six years of the CB project (2005–2011), 175 South Asian Water (SAWA) fellowships were awarded to students for research theses for their master's (ME and MSc), MPhils and PhDs in integrated water resource management throughout South Asia. Eighty per cent of the fellowships were awarded to women researchers, and 20 per cent to researchers from lower socio-economic groups and deserving male students. Many Fellows have already got employment in the water sector and are contributing their bit with a larger vision of water. In some instances, students' research has directly contributed in community interventions to solve water problems with a strong gender and equity approach, including actively supporting the marginalised in a legal water pollution dispute. For instance, a Senior Professor at the Centre for Water Resources, Anna University, Chennai, India (now retired), said,

> For the very first time in our departmental history, we were approached by these communities (living on the fringes of the local water body) to allow them to use the technical data in a court hearing against the municipality (relating to the rampant dumping of wastes, including hospital and industrial wastes). It was the first time that we felt our work had such a direct, immediate relevance for society.

4.3.3 Research

The project developed research programmes and methodology with the universities that were impact-oriented, societal-driven and participatory in approach. Thus, all the universities, through their PhD and master's students conducted research on themes that addressed current problems in water resource management in their selected study areas. Furthermore, the research process was conducted in a participatory manner involving primary stakeholders. The research themes developed included sustainable management of land, water and ecological resources (Bangladesh); the impact of urbanisation on water resources, and the availability and quality of water in and around Chennai (India); institutional innovation for sustainable water supply and sanitation services in the upper Mahaweli watershed (Sri Lanka); and integrated river basin management (Nepal). The research proposals were all linked to ongoing projects, thus ensuring that the research was embedded in society and address real problems (Prakash et al., 2009). The trainings on the various subjects/issues as well as the training on field research methodologies were also geared towards building the capacity of the faculties and students towards successfully conducting such research.

4.3.4 Advocacy

As stated earlier, the project's major focus is on IWRM and gender and water, both poorly researched areas that call for supportive activities and strong advocacy support. Especially in the case of gender and water, past experience indicated that a multi-pronged effort is needed, with capacity building and advocacy backed with empirical studies and appropriate material for dissemination in order to facilitate gender-responsive changes in the water sector. An advocacy effort based on a clear understanding of the impact and potential contribution of women to the water sector can promote gender equity as well as pinpointing further gaps in research and education programmes.

As a part of this effort, the project carried out a concerted upgrading of professional's skills and expertise with special focus on WWPs. Furthermore, training and sensitising male and female water professionals on gender and equity issues in water management in South Asia were also conducted. The bulk of the fellowships for master's and PhD in the intervening universities were reserved for women applicants to create a critical mass of women professionals at the regional level. Two specific aspects of this component included to enhance the status of WWPs in South Asia and set an agenda for networking for more gender-sensitive water management in the region.

The project recognised that in South Asia though many female water professionals are already active in the sector, their visibility is low and they are also constrained at lower levels in their institutions, as was seen for the government water departments too. Their opportunities for advancement and capacity building are limited due to organisational cultures as well as the prevalent social norms. The selected activities of the project thus focused on a better understanding of their current status, recognition of their professional needs and also give expression to their aspirations. It also raised their awareness on the present and potential contribution of women as activists and professionals within the current cultural constraints of South Asia. The increasing interest on the question of gender and water has emanated from the women's movements and struggles. In the past, this has compelled the state to address some inequities and disadvantages that women face. However, there is little attention given to theoretical debates in the water sector and hence most of the efforts at making the water sector inclusive become mere lip service or remain partially addressed. This is evident from the ad hoc introduction of programmes in the sector, which seek women's participation in community-based domestic water programmes but do not result in empowerment. Moreover, most of these programmes remain focused on the poor rural women and their collectives to regenerate and effectively manage the resource. Little research has been done in the area of the professionals who are in the key decision-making or implementing positions in the organisations in both the bureaucracy and other institutions and how their actions or inactions set the agenda for policy and programme in the water sector. Furthermore, there is little thought given to the women who work as water professionals in different water subsectors and their constraints and positive influences on the sector as a whole. By WWPs, we simply imply women who are employed in the water sector in different capacities. It is in this context that a study on WWPs becomes critical. While reiterating the need for a much deeper theoretical treatment of the question, it was felt that there is a need to look at women at various levels in the water sector.

Therefore, the CB project initiated an exploratory study at a South Asia level in Sri Lanka, Nepal, India and Bangladesh primarily to understand the profiles, numbers and constraints of WWPs in the region (SaciWATERs and SOPPECOM, 2011). The focus of the study was to understand the participation rate of women professionals in water sector and why they are so few. What are their concerns and does their presence make any difference to the gender mainstreaming agenda? Does reform in policy bring any more visibility to their concerns, does it provide them any more space than it did in the past and does this space lead to fruitful outcomes in terms of gender equality? These and other related questions define the scope of the study. The major findings of the study were:

1. *Low numbers:* All of the countries show that a very small number of women (ranging from 1.5 per cent to 17.4 per cent) are working as WWPs in South Asia. Therefore, there definitely is a glass ceiling for women in the sector, as during till the time of the study in 2009, there were no women at the top level in any of the departments dealing with water in South Asia.
2. *Stereotyping:* The study reports four ways of stereotyping: (a) women are relegated to administrative and desk work of varying nature and not encouraged in site-based work and financial transactions solely on the basis of their gender. (b) Men find it difficult to accept women in leadership roles and it is believed that women make better subordinates as men felt more comfortable in patronising them in brotherly/fatherly manner. (c) Any description of a good water sector officer started with a 'he' that assumes that water professional will be a man rather than a woman! It is almost always seen as men possessing the qualities of being able to think rationally and to take instant decisions—qualities seen as required for a good and successful water sector officer. (d) Women are expected to behave and dress in certain ways, for example, they should not talk or laugh too loudly, should not be dynamic or aggressive, should not apply make-up in the office, should wear traditional feminine clothing, etc. These gender stereotypes were extended to women working in government departments dealing with water issues and hence on a broader sense, restricted women's mobility in the field or participating actively or giving opinions representing women's issues.
3. *Gender neutrality with seniority:* Across all the countries it was found that as women moved up the seniority ladder, people's attitude towards them was more neutral and staff felt that 'gender should not come into organisation'.
4. *Absence of basic facilities:* In all the countries, there is clearly an absence of basic sanitation facilities like clean and separate toilets

for women in the workplace. There is also a lack of other facilities such as childcare/crèche facilities, transport and accommodation during field work, and adequate maternity leave.

5. *Hesitant but changing profile of the water bureaucracy:* Changes from the ethnocentrism towards more multidisciplinary and interdisciplinary approaches are being introduced in small and cautious ways in all the countries.

4.3.5 Training in Gender, Water and Equity for Working Water Professionals

The project recognised that policies, projects, programmes and researches in South Asia have attempted to integrate women, with varying degrees of success, in countering the multiple processes that contribute to inequalities. However, not everyone is equipped to understand the complexity of the water sector as it relates to gender and concerns of equity. In this context, trainings for working professionals in South Asia were conducted. The objectives of the training programmes were to (a) understand the larger political context of water distribution and reforms; (b) strengthen perspectives on gender, water and equity issues; and (c) examine analytical frameworks that could incorporate gender in the planning and implementation of programmes in the water sector. The project trained 120 working professionals across South Asia equipping them with concerns on incorporating gender and equity issues in research, action and programming.

4.4 Conclusion

This chapter documented the approaches, outputs and outcomes of a unique initiative that was operational in four South Asian countries called the 'Crossing Boundaries' during 2005–2011. The project focused on education, impact-oriented research, networking and advocacy as combined effort to contribute to a paradigm shift in water resources management in South Asia. This chapter presented the regional, collaborative, partnership-based capacity-building initiatives undertaken by the project. This section outlines some of the outcomes, learning and challenges in carrying out the programme in three areas—improved understanding of interdisciplinary and participatory approaches, influencing problem-oriented research and creating cadre of water professions with an alternative vision as long-term strategy to make the sector more gender-sensitive.

First, the interdisciplinary teaching and capacity building has enabled students to pose research questions in an interdisciplinary manner, which has a societal relevance. Apart from receiving interdisciplinary courses in the classroom, students also underwent various training programmes on interdisciplinary research. Previously, the research questions were more multidisciplinary than interdisciplinary in nature. The multidisciplinary questions were non-integrative in nature that represented a mixture of disciplines. Each discipline, therefore, retained its methodologies and assumptions without change or development from other disciplines within the multidisciplinary relationship. Changing to an interdisciplinary field of study helped in crossing traditional boundaries between academic disciplines or schools of thought, as new needs and professions emerged catering to the real-life questions and seeking practical answers to the questions asked.

The first set of activities employed during the initial implementation period of the project was to reshape the technically oriented water resources management curriculum of the partner institutions into an interdisciplinary programme through introduction of three new courses, namely Field Research Methodology, IWRM and Gender and Water. To introduce these courses, trainings were conducted for the existing faculty as they were not exposed to tackling interdisciplinary courses because their own training was mostly on civil engineering, which was devoid of interdisciplinary subjects. These training activities have sensitised staff to think beyond technical solutions in addressing water resources management issues. The academic staff, especially the younger group, has very little exposure to societal issues since they have been conducting research within narrow disciplinary areas. Continuous exposure to training programmes on societal issues and events which address water resources management problems in an interdisciplinary manner has re-oriented technically qualified academic staff to modify traditional engineering curriculum that are engaged in teaching.

Second, the research programmes and methodology focussed on impact-oriented, societally driven and participatory approaches. Thus, all the partner institutions through their PhD and master's students have conducted research on themes that address current problems in water resource management in their own selected study areas. The research process is conducted in a participatory manner involving primary stakeholders who had formed the agenda for research. The trainings on the various subjects as well as the training on field research methodologies were also geared towards building the capacity of the faculties and students to successfully conduct such research. As a result, the problems were conceptualised differently and as a process that reflected the ground realities of water management problems.

Third, supporting the new cadre of young and promising water professionals with additional skills of interdisciplinarity, gender and participatory approaches, and strong networking of WWPs led to gender-sensitive research and action. This was done at two levels: at the student's level and at the working water professional's level. The programme, thus, raised awareness about socio-economic, gender, political and environmental issues in water resource management among water professionals who have been trained.

One of the challenges of this project was the institutionalisation and appreciation of interdisciplinary approaches in technical education. The problem-based interdisciplinary research was new to many partner institutions. The traditional practice is to guide the students through their respective supervisors for a specific research project with clear objectives. In this case, academic staff along with the students addressed a common theme from different angles to address a major water management problem identified by the universities. However, apart from the key learning, introducing an interdisciplinary education was challenging in the rigid university education in South Asia. In the process, faculty found it difficult to seek an interdisciplinary base for research when their own training was within demarcated disciplines. Therefore, a major learning of the project was that the content of interdisciplinary teaching needs to be grounded in the reality of the sector. Through an initial course in field work, the students get to know the ground reality not only at the intellectual level, but even at the experiential level. The students were encouraged to approach theory in the light of practice. This made students willing to seek impact-oriented research.

If given proper support in terms of specially designed, taught and self-study courses, there is greater chance for high-quality research, not only in academic terms but also in terms of its relevance and utility to the outside stakeholders. At the same time, because of the same factors, the students are less inclined and less capable due to lack of knowledge, awareness and tools for interdisciplinary research.

Note

1. The six partner institutions are: Centre for Water Resources (CWR), Anna University, Chennai, India; Postgraduate Institute of Agriculture, University of Peradeniya, Kandy, Sri Lanka; Bangladesh Centre for Advanced Studies (BCAS) in collaboration with Institute for Water and Flood Management; Bangladesh University of Engineering Technology (BUET), Dhaka, Bangladesh; Nepal Engineering College (NEC), Katmandu, Nepal and Tata Institute of Social Sciences (TISS), Mumbai, India.

References

Ahmed, Sara. 2005. *Flowing Upstream: Empowering Women through Water Management Initiatives in India*. Ahmedabad, India: Centre for Environment Education and New Delhi: Foundation Books.

Cap-Net, India, SOPPECOM, Women Water Network (WWN). 2007. *Water for Livelihoods: A Gender Perspective*. (Based on writings of S. Ahmed, S. Arya, K.J. Joy, S. Kulkarni, S.M. Panda, and S. Paranjape) Ch. 5, pp. 60–76. Pune: Cap-Net, India, SOPPECOM and WWN.

Joshi, D. 2005. Misunderstanding gender in water: Addressing or reproducing exclusion, in A. Coles and T. Wallace (eds), *Gender, Water and Development*, Ch 8. London, UK: Bloomsbury Academic Publication.

Joshi, Deepa and Margreet Zwarteveen. 2011. Gender in drinking water and sanitation: An introduction, in Zwarteveen, Margreet, Sara Ahmed and Suman Gautam Rimal (eds), *Diverting the Flow: Gender Equity and Water in South Asia*, pp. 161–174. India: Zubaan-SaciWATERs.

Joshi, Deepa. 2011. Caste, gender and the rhetoric of reform in India's drinking water sector. *Economic and Political Weekly*, xlvi(18): 56–63.

Kapadia, Karin (ed). 2002. *The Violence of Development: The Politics of Identity, Gender and Social inequalities in India*. New Delhi: Kali for Women.

Krishnaraj, Maithreyi. 2011. Women and water: issues of gender, caste, class and insutitions. *Economic and Political Weekly*, xlvi(18): 37–39.

Kulkarni, Seema. 2012. Situation analysis of women water professionals in South Asia, in Krishna, Sumi and Arpita De (eds), *Women Water Professionals: Inspiring Stories from South Asia,* pp. 201–247. New Delhi: Zubaan and SaciWATERs.

Lahiri-Dutt, Kuntala (ed). 2006. *Fluid Bonds: Views on Gender and Water*. Calcutta, India: *Stree*.

Mollinga, P. 2009. *Strengthening IWRM Education in South Asia: Which Boundaries to Cross?* Hyderabad, India: SaciWATERs.

Prakash, Anjal, Edwin Rap and Dinis Juizo. 2009. Influencing change by bridging disciplinary divides: Learning from three regional capacity building networks on Integrated Water Resources Management. Presentation at 5th World Water Forum 6.1.3 Session on Knowledge Development and Capacity Building. Istanbul. March 16–22.

Zwarteveen, Margreet, Sara Ahmed and Suman Gautam Rimal (eds). 2012. *Diverting the Flow: Gender Equity and Water in South Asia*. New Delhi: Zubaan-SaciWATERs.

5

Gender and WASH: Capacity-building Initiatives

Swati Sinha

5.1 Introduction

Gender refers to socially constructed roles, behaviour, activities and attributes that a particular society considers appropriate and ascribes to men and women. These distinct roles and the relations between them may give rise to gender inequalities where one group is symmetrically favoured and holds advantage over another (WSP, 2010). The need to incorporate gender perspectives in water and sanitation programmes has been asserted by development practitioners in national and international forums. It has been observed that women and girls are most often the primary users, providers and managers of water in households and are the guardians of household hygiene (WSP, 2012: 9). In the absence of proper systems of water access, they are forced to travel long distances to fetch water leading to waste of time and energy. Improved access to water and sanitation services would benefit women the most, contribute to redress gender inequality, positively impacting socio-political and economic position of women, improve health and security of women and their family, and free them to engage in social, economic and political activities (WSP, 2012: 9). These have been reinforced by principle number 3 of Dublin Principle that stresses on the need to empower women to participate at all levels of water management according to their guiding principle that states that 'women play a central part in the provision, management and safeguarding of water' (Fong et al., 1996: 2–3). This chapter is divided into six sections. I begin with establishing linkages between gender and WASH and explaining the purpose of the study and methodology. The second section demonstrates the importance of capacity building in incorporating gender and WASH. The third section deals with the information related to existing structures and institutions for capacity-building initiatives in WASH in India. The fourth section presents the analysis of select capacity development materials from government and non-governmental organisations (NGOs). Based on the analysis of the reviewed documents, the fifth section discusses the measures to include gender into overall capacity-building strategy in WASH in India. The concluding section includes recommendations and makes a contribution towards gender-inclusive WASH services.

5.2 Linkages between Gender and WASH

The United Nation's Millennium Development Goals (MDGs) are an internationally agreed development framework. Among the eight goals listed include 'to halve by 2015, the proportion of people without sustainable access to safe drinking water and basic sanitation'. This is directly linked with WASH (Millennium Project, 2006). Its inclusive approach is strongly related to gender mainstreaming in WASH. Access to WASH would give impetus in fulfilling MDG 3, 'Promote gender equality and empower women' and Target 4 of MDG 3, 'eliminate gender disparity in primary and secondary education preferably by 2005 and in all levels of education no later than 2015' (Millennium Project, 2006). This is based on the fact that school-going young and adolescent girls assist their mothers in collecting water from far-off places, taking up a considerable part of their day leading to high frequency of absenteeism in school or not attending school at all (UNDESA, 2005). Another reason for absenteeism in schools among adolescent girls is the lack of provision of adequate sanitation in schools. Due to absence of separate toilets for girls in school, most of the adolescent girls do not go to school during menstruation (Mahon and Fernandes, 2010). Absence of good sanitation and hygiene and ignoring the issues related to menstrual management in latrine design contribute to absenteeism and dropout of girls (Bhardwaj et al., 2004: 4).

Furthermore, a report submitted by Indian Council of Medical Research (ICMR) on the estimates of maternal mortality points out unhygienic living conditions as one of the reasons for health problems of mothers and newborns, leading to maternal and infant deaths (Indian Council of Medical Research, 2003: 1). This clearly indicates the need for safe and adequate sanitation to achieve MDGs 4 and 5 to 'reduce child mortality' and 'improve maternal health,' respectively.

5.3 Purpose and Methodology

This chapter reviews different capacity-building programmes on WASH presently operational by Government of India and Indian NGOs from gender and equity lens. The main objective is to identify the gender gaps in capacity-building initiatives in WASH-related issues. The study aims at drawing attention of policy makers, researchers, thinkers and civil society representatives towards the need for concentrated efforts for gender mainstreaming in capacity-building initiatives in WASH. I also aim to find out mechanisms for gender inclusion in capacity-building initiatives for WASH. This chapter points out the relevance of gender and WASH in global context by interlinking women and WASH with the international agenda namely 'Dublin Statement' and MDG's goals and targets. It highlights the key role of capacity-building initiatives by government and NGOs in incorporating gender into WASH.

For the study and analysis of gender components in capacity-building initiatives in WASH, effort was made to get updates on existing capacity-building programmes of the government from the official websites of Department of Drinking Water and Sanitation (DDWS) in India. Secondary literature review included gender mainstreaming in WASH and ongoing initiatives for inclusive approach in WASH sector at national and international levels. Select policy documents, training reports, training modules, brochures, Information, Education and Communication (IEC) materials and case studies prepared by the government and NGOs were collected and reviewed. In addition, I have reviewed select training reports, manuals, modules and materials.

5.4 Capacity Building, Gender and WASH

The United Nations Development Program (UNDP) has defined 'capacity' as 'the ability of individuals, institutions and societies to perform functions, solve problems, and set and achieve objectives in a sustainable manner'. Capacity development is the process through which the abilities to do so are obtained, strengthened, adapted and maintained over time (UNDP, 2006). UN Water (2006) explains building capacity as bringing together more resources, more people (both women and men) and more skills. It looks like that in developing countries, capacity-building efforts for water supply and sanitation are focused more on water as a resource targeted towards technical water supply specialists. It states:

> very few programmes and projects are aimed at expertise in social development, sanitation, or hygiene education that emphasizes a gradual scaling down to those responsible for operation and maintenance of water supply and sanitation, who are primarily women. Targeting women for training and capacity building are critical to the sustainability of water and sanitation initiatives, particularly in technical and managerial roles to ensure their presence in the decision-making process. (UN Water, 2006: 6)

Therefore, stronger linkages emerge to build capacity of communities, implementing agencies such as government and NGOs with regard to gender issues in WASH. These links are established both from empirical studies as well as in policy documents. The need for capacity development among stakeholders and recommendation to improve women's access to water resources through enabling development and gender mainstreaming is strongly emphasised in many policy documents (IFAD, 2007).

> Significant support and capacity-development efforts are required to enhance the participation of rural women in decision-making processes for water management. Training and capacity development among women to enable them to take up leadership roles, to voice their concerns without any hesitation and to enhance their technical skills are essential if the benefits of water projects in reducing poverty and improving livelihoods are to be equally distributed. Also, rural men need to be engaged in empowering rural women, particularly in societies where the support of men for such initiatives is required. (IFAD, 2007: 23)

5.5 Institutional Set-up for Capacity-building Initiatives of Government of India

NRDWP and NBA—MDWS, formerly under the Ministry of Rural Development (MORD) as DDWS, is a nodal department for the overall policy planning funding and coordination of programmes of drinking water and sanitation in the country (MDWS, 2011a). In rural areas, it runs two major programmes named as 'National Rural Drinking Water Programme' (NRDWP) and *Nirmal Bharat Abhiyan* (NBA). The Government of India's effective role in the rural drinking water supply sector started in 1972–1973 with the launch of Accelerated Rural Water Supply Programme (ARWSP). In the year 1991–1992, the programme was renamed as Rajiv Gandhi National Drinking Water Mission (RGNDWM) that puts stress on water quality, appropriate technology intervention, human resource development (HRD) support, and other related activities (GOI, 2010: iii). RGNDWM has modified the existing rural water supply guidelines, made effective from 1 April 2009 as the National Rural Drinking Water Programme (NRDWP, n.d.). It emphasises on ensuring sustainability of water availability in terms of portability, adequacy, convenience, affordability and equity while also adopting decentralised approach involving *Panchayati Raj* Institutions (PRIs) and community organisations (GOI, 2010). For sanitation, the government started the Central Rural Sanitation Programme (CRSP) in 1986 primarily with the objective of improving the quality of life of the rural people and also to provide privacy and dignity to women. The concept of sanitation was expanded to include personal hygiene, home sanitation, safe water, garbage disposal and waste water disposal. With this broader concept of sanitation, CRSP adopted a 'demand-driven' approach with the name 'Total Sanitation Campaign' (TSC) with effect from 1999. The revised approach emphasised more on IEC, HRD and capacity development activities to increase awareness among the rural people and generation of demand for sanitary facilities. To give a fillip to the TSC, the Government of India also launched the *Nirmal Gram Puraskar* (NGP) that sought to recognise the achievements and efforts made in ensuring full sanitation coverage. The award gained immense popularity and contributed effectively in bringing about a movement in the community for attaining the *Nirmal* status thereby significantly adding to the achievements made for increasing the sanitation coverage in the rural areas of the country. Encouraged by the success of NGP, the TSC is being renamed as '*Nirmal Bharat Abhiyan*' (MDWS, 2013). As per the guidelines for NRDWP, the national goal of the programme is to 'provide rural persons with adequate safe water for drinking, cooking and other domestic basic needs on a sustainable basis. The basic requirement should meet minimum water quality standards and be readily and conveniently accessible at all times and in all situations'

(MDWS, 2011b). As per the recent guidelines of NBA, the main objectives of the programme includes (1) accelerating sanitation coverage, (2) ecologically safe and sustainable sanitation, (3) community-managed environmental sanitation and (4) motivating communities and PRIs and promoting sustainable sanitation facilities through awareness creation and health education (MDWS, 2011c). Guidelines of the NRDWP and NBA reflect on the need for stronger capacity-building component in WASH. It has envisaged setting up key resource centres (KRCs) at national level and Communication and Capacity Development Units (CCDUs) in each state.

5.6 Key Resource Centre

With the new approach of the NRDWP programme, there is a need felt for 'building a multi-level cadre of motivated, skilled and trained personnel in the rural water supply sector'. There is a need to identify KRCs at the national level 'with high repute and experience engaged in working on these issues, in imparting training and in other activities to build capacities of different stakeholders in drinking water and sanitation sector'. National KRCs are responsible for organising capacity-building programme to government functionaries at various levels on all issues pertaining to drinking water and sanitation. Analysis of the objectives of KRCs includes capacity building of Public Health Engineering Department (PHED) engineers, PRI representatives, master trainers and other stakeholders so as to enable them to carry out their responsibilities, intellectually and professionally in an effective and sustainable manner and foster respect for rural community's rights. Besides, KRCs also aim to enhance the capacity of CCDU (MDWS, 2011d).

5.7 Communication and Capacity Development Unit

The main objective of setting up of a CCDU in each state is to 'develop state-specific information, education and communication strategy for reform initiatives in water and sanitation and to provide capacity development of functionaries at all levels'. The guidelines explain that the key function of CCDU is to 'provide IEC/HRD support to State Water and Sanitation Mission, provide IEC/HRD inputs to all ARWSP, Swajaldhara and TSC projects in the state, and documentation and dissemination of successful IEC/HRD initiatives taken within the state and in other states and agencies'. The list of activities under IEC and HRD focuses on training PRIs, NGOs, district-level officers and state-level functionaries, and school teachers on subjects relating to technological options, water quality issues, water quality monitoring, sanitation and hygiene, social mobilisation, and school sanitation (MDWS, 2011e).

5.8 Capacity-building Intervention on Gender and WASH

This section reviews the coverage of gender components in policy documents and strategy papers, gender representations in trainings, gender-specific contents such as menstrual hygiene management, gender mainstreaming, gender needs for WASH services, gender stereotyping and reinforcing patriarchal norms in training documents. It examines how far these issues are covered in the regular training programme and curriculum. The summary is as follows.

- Twelfth Five-year Plan, 2012–2017: The Report of the Working Group on Rural Domestic Water and Sanitation (Planning Commission, 2011) uses the term 'gender', 'gender equality' and 'gender issues' on several occasions and highlights the need to incorporate gender issues in water and sanitation projects (Planning Commission, 2011). It has elaborately mentioned the role of civil society in addressing the issues related to menstrual hygiene management (Planning Commission, 2011: 118). Provision of girl-friendly toilets with suitable water facilities and for safe menstrual hygiene management in all schools and to make sanitary napkins available to them in rural schools either free of cost or at affordable prices have also been recommended (Planning Commission, 2011: 208). However, training documents from both government and NGOs show that specialised focus on gender issues is missing. The policy documents are not supported by evidence and remain written words only. Menstrual hygiene management is an important part of WASH and comes as one of the major components of sanitation. Preliminary analysis done for the study shows that it has not been a part of select training manuals, modules and handbooks.

- An *Information guide to Village Panchayat Presidents on Drinking Water and Sanitation* (CCDU, 2007) focuses on roles and responsibilities on Village Water Supply and Sanitation Committee (VWSC) in the provision of safe drinking water and promoting WASH. Though this document recognises women as primary stakeholders and as contributing towards educating the mass and participating in the community meeting on water and sanitation, the specific role of women members of VWSC have not been mentioned.
- Report 'Swajaldhara and TSC training on Capacity-Building Programme for Indian State, 2005', prepared by KRC, Centre for Good Governance and Uttaranchal Academy of Administration (GOI, 2005), presents a detailed information on training themes, target group and the geographical coverage during May to November 2005. In Phase 1 of Swajaldhara and TSC programme, training was imparted on community participation and the applicable components for Swajaldhara, observation study tour for community-managed water and sanitation and water and sanitation hygiene promotion strategy. The document provides detailed information on representations of the officials from various states and regions, which is mainly government offcials at various levels from different states. There exists information on profile of resource persons, content of the training selected for the capacity-building programme, methodology adopted for designing, and imparting training programmes and gender-wise representation in the training programme.

However, critically analysing reports from gender and equity lens, it is clear that representation of women as participants of the programme is very low. Out of total 433 participants from various states and regions, number of women participants is only 30, which is only 6.9 per cent of the total participants. In the resource person category, there is only one female resource person out of 15 resource persons from the Center for Good Governance and Uttaranchal

Academy of Administration. Overall, the training content points towards the need for more emphasis on the participation of women in water management and also a need to be actively involved in all programmes at the decision-making level particularly at community level. The training programme promoted the participation of women in water management as emphasised under Swajaldhara programme. The guidelines emphasise active participation and involvement of women in 'planning, choice of technologies, location of systems, implementation, operation and maintenance of water supply schemes' (PIB, 2005). As mentioned earlier, gender representation in training is low and a matter of concern when we talk about and focus on active participation of women in water management at the community level. However, there is the need of gender balance in water management structures from top to bottom for programmes to be successful. Women should be represented in water bureacacy at the national, state, district and block levels.

- A technical note series of RGNDWM (2004) on 'TSC and Capacity Building: Activites and Modules' suggests 7 modules for stakeholders on different sanitation-related themes. These focus on hygiene education, school sanitation, construction of toilets and marketing of sanitary marts, safe disposal of excreta, etc. The module has a session relating to access to sanitaton for ensuring women's dgnity, need for women sanitary cum bathing complex, and the construction of separate toilets in schools for girls. In spite of its focus on hygiene education, there is no session separataly dealing with water and sanitation-related issues of women.
- Pictorial guide on hygiene education produced by MDWS (2011f) has illustrated personal hygiene in many ways. School-going boys and girls are shown performing activities like brushing the teeth, bathing, brushing their hair, using soaps to wash hands after defecation, and before start eating, etc. It has also talked about the immunisation and vaccination calendar and food that needs to be fed to the newborn and infants. This document conveys the message on sanitation and hygiene behaviour effectively, but promotes gender stereotyping and male child preference. For instance, a picture displayed on page no. 3 shows a girl child pouring water to enable a boy child to wash his hand. Similarly, a picture on

page no. 11 shows the type of nutrition s/he should provide as the child grows from 0 months to 12 months. Minute observation reveals that the mother is all about nurturing a male child and not a female child. In the current scenario when the figures of female foeticide are quite alarming in India, there is a need to take extra care to avoid these in IEC materials.

- 'Sanitation and hygiene advocacy communication framework 2012–2017' that UNICEF prepared for MDWS is an important document on hygiene education and behaviour change. Its advocacy and communication strategy focuses on four critical sanitation and hygiene behaviours particularly building and use of toilets, safe disposal of child faeces, hand washing with soap after defecation, before food and after handling child faeces, and safe storage and handling of drinking water. The document advocates for strong capacity-building component. Therefore, it proposed for implementation framework (UNICEF, 2012: 29) at state and district levels for effective execution of the strategy and includes training module development on sanitation and hygiene issue and training of stakeholders as preparatory activities for implementation plan. However, it is unfortunate that it has totally ignored the gender issues and has not sufficiently covered menstrual hygiene management. The document is comprehensive in detailing out who, what and how of advocacy and communication strategy. Therefore, the inclusion of menstrual hygiene management as one of the critical sanitation and hygiene behaviours in advocacy strategy and communication framework would certainly encourage incorporating the component in implementation plan.

5.9 Capacity-building Intervention by NGOs and International Agencies

United Nation's documents on Best Practices in Gender and WASH in various countries (Office of the Special Adviser on Gender Issue and Advancement of Women, 2006) have projected capacity building as one of the important components for gender mainstreaming in WASH sector.

Importance of capacity-building intervention to mainstream gender in WASH has equally been recognised by NGOs and international agencies. As a result, NGOs working in WASH or water sector has initiated capacity-building projects within its policy framework. There are international, national and regional level not-for-profit agencies in India that work on issues related to gender and WASH. WaterAid India, SaciWATERs and Gramalaya are some of the organisations amongst others in India that have initiated capacity building of stakeholders on gender and WASH. This section analyses the strategies and outcomes for promoting gender inclusion in WASH.

A case of community sanitation project in slums of Tiruchirappalli district of Tamil Nadu state in Southern India, entitled, 'India: From Alienation to an Empowered Community—Applying a Gender Mainstreaming approach to a Sanitation Project Tamil Nadu' (Victor, 2006) is quite relevant in the context of understanding the role of capacity building in gender mainstreaming in WASH. This is the story of non-functional status of dry latrines in the slums contributing to filthy environment and causing in the increase of the disease burden on the inhabitants of the area. It was revealed that male community leaders and the government were not doing anything to resolve the issue. Taking stock of the poor sanitation status in the area, Gramalaya intervened for the 'installation of drinking water facilities and individual toilets, as well as community mobilisation with a focus on gender mainstreaming'. As a result, the toilets were converted into pay and use toilet integrating vermi-composting with the sanitation plan. The NGO also constructed child-friendly toilets and educated women, men and children on hygiene practices. Women's empowerment approach and organising women in the form of women's self-help groups and their capacity building in accounting and accessing government schemes resulted in a successful project.

The project design focussed on the formation of women's self-help groups for the primary community mobilisation and sanitation management mechanism, using them to establish a savings and credit scheme. It also built capacity of women's self-help groups in accounting and accessing finances and services from the government (Victor, 2006: 81–87).

SaciWATERs is a policy research institute. This organisation has been organising gender and equity trainings at the South Asia level with support from Dutch Government, Cap-Net and Gender and Water Alliance.

Since 2009, they have imparted five training programmes in South Asia with the aim of gender and equity in domestic water and sanitation and irrigation sector. This is the only up-to-date training programme available that is specially dealing with gender issues. However, it is not dedicated for WASH as it covers a larger domain of gender in Integrated Water Resource Management Programme and covers South Asia (SaciWATERs, 2011).

5.10 Agenda for Future Action: How to Include Gender Issues in WASH-related Capacity-building Programme?

Review of the guidelines and the select material clearly reveals that though gender issues get prominence in policy documents, it has not translated into effective capacity-building programmes, curriculum and training materials for mainstreaming gender into WASH. Some important components such as menstrual hygiene and management are totally missing from most of the WASH-related trainings. We suggest the following measures to integrate gender into capacity-building initiatives in WASH in India.

5.11 Need for Gender Mainstreaming in WASH

In water and sanitation projects, the mainstreaming of gender can be done through understanding the differences in needs and priorities of women, men, girls and boys that arise from their different activities and responsibilities; and analysing the inequalities in access to and control over water resources and access to sanitation services (UNICEF, 2003a). The need for gender mainstreaming in WASH is based on the analysis that, 'Women's presence is critical to the sustainability of water and sanitation initiatives, particularly in technical and managerial roles to ensure they contribute in decision-making processes. Ensuring women gain access to information about project plans and resource allocations is also essential (WaterAid, n.d.). It has been argued that gender mainstreaming in the sector has largely about women making projects work, rather than projects working to reduce gender inequities (Joshi and Zwarteveen, 2012: 161). Against this backdrop, a strategy for capacity building that can incorporate such elements as redressing gender inequalities and addressing practical gender needs as well as strategic gender needs, which empower them to participate in decision-making levels is a critical gap at present.

5.12 Gender-inclusive Capacity-building Strategy

Capacity building should not only be limited to structured training but also incorporate other aspects like awareness creation, education, sensitisation, exposure to best practices and mentoring. It should not be a one-time activity but an ongoing process and serve as a sustainable vehicle for social change. It is recommended that the government and civil society and other stakeholders must work towards comprehensive gender-inclusive capacity-building strategy in WASH. It should include the following:

- Conduct gender analysis and find out gender issues in WASH. This could be related to existing gender inequity in access to WASH services, lack of awareness about the existing village level institutions for water and sanitation services, poor status of menstrual hygiene management, ignorance and lack of participation of women in water and sanitation-related programmes.
- Undertake capacity building needs assessment. This should incorporate learning needs of the community men and women, representatives of village-level institutions dealing with WASH, government officials at district and state levels from rural water supply, and sanitation department on gender perspectives.
- In the capacity-building programme in WASH, gender should be incorporated right from the designing of the training programme, deciding the content, the resource person, the methodology and monitoring and evaluation of the programme.
- It is recommended to integrate key points to check gender equity developed by UNICEF (2003b). This can be used in capacity-building initiatives on WASH at different levels to find out different needs, interests and priorities of women, men, girls and boys;

to use gender perspectives to gather information; investigate the gender issues particularly gender gaps; barriers and immediate and underlying causes; examine gender balance in programme objective; inclusion of physical and cultural aspect of gender, explore gender balance in decision-making, gender-specific elements in water and sanitation strategies, examine impact related to gender inequalities and measure to monitor for separate effects.

- It is important to review existing training and capacity-building programmes from gender and equity lens. A much more detailed analysis and structured review of training curriculum is needed to examine gender-related components. This would include the preparation of an exhaustive list of training curriculum, and capacity-building strategies of both the government and NGOs. This should be followed by finding out key gender gaps and setting up some of the gender-sensitive indicators. Thorough examination from gender lens would help in finding gender gaps in the existing capacity-building programme.
- Documenting best practices on gender mainstreaming in WASH is very important to make an effective capacity-building strategy. That would become a base to find out key areas where gender needs to be incorporated and what are the successful approaches that could be replicated in other areas.

5.13 Tools for Gender-inclusive Capacity-building Programme

Use of appropriate tools and techniques for capacity building has an important role in sensitising stakeholders on gender and WASH issues. There are examples of using participatory approaches in capacity-building programmes for gender sensitisations in development programmes. Use of participatory methodology in the training on gender sensitisation to understand the framework of gender analysis, organised by Haryana Forest Department as part of Haryana community forest project is one such example (Ray, 2000). Participatory training is a learning process, which involves learning of new skills, concepts and behaviour. It facilitates learning and critical thinking process to create an enabling environment (PRIA, 2002: 11). My field-level experience in participatory training on development issues and gender and WASH-related projects in various parts of the country suggests some of the tools to mobilise greater participation from women. These include conducting informal interaction and meeting with the target group particularly women on frequent intervals, organising semi-structure meetings with the target group to build their perspectives, promoting learning by organising exposure visits and exchange visits, organising campaign on issues related to gender and WASH, involving local groups of both men and women to develop IEC materials, strong information dissemination, model building and so on.

5.14 Other Components to Make Capacity-building Programme Gender-inclusive

There are factors that hinder participation of women in structured training programmes on WASH-related issues organised by the government or local civil society organisations (CSOs). My engagements in community-level capacity-building programmes reveal that women rarely participate in such programmes. Key reasons for lack of participation include household responsibilities, lack of time, lack of provision for safety and security, lack of motivation, lack of encouragement to participate in such programmes by family members, etc. Therefore, there is a need to check with the target group about their suitability of venue and time selected for the training programme, ensure security and safety of women at the training venue to avoid incidences related to eve teasing or sexual harassment. Also, one needs to take the family members into confidence ensuring their safety and security at the training venue, make arrangements of other needs for women such as arrangement for baby care centre, and provide handholding support for women trainees attending training with small child. There is a need to observe gender sensitivity of the trainers irrespective of gender and gender-sensitive training contents. Further it is important to encourage community members to break social norms and stereotyping, etc.

5.15 Conclusion

Gender is an integral part in water and sanitation-related discourse in India and worldwide. It has widely been realised that improved access to water and sanitation services would benefit women most and redress gender inequality. Access to WASH would give impetus in fulfilling MDG 3, and 'eliminate gender disparity in primary and secondary education preferably by 2005 and in all levels of education no later than 2015' (Millennium Project, 2006). Building the capacities of all stakeholders on gender in WASH-related issues is important for integrating gender in WASH and for the sustainability of water and sanitation initiatives. Existing provisions to set up KRCs and CCDUs under NRDWP and NBA are an effort to integrate capacity-building component into the programme. The capacity-building objectives of KRCs focus on building knowledge, enhancing skills and changing attitude of the target groups.

The main objective of setting up of a CCDU in each state is to 'develop state-specific information, education and communication strategy for reform initiatives in water and sanitation and to provide capacity development of functionaries at all levels'. Integration of WASH and gender into capacity-building programme is a means to an end. Therefore, the need to scan select training resource materials, including IEC materials, training modules used by the KRCs, CCDU, government training reports and other agencies with a gender and equity lens. It was found that the specialised focus on gender issue is missing. Preliminary analysis shows that menstrual hygiene management has not been a part of select training manuals, modules and handbooks. Additionally, there is lack of gender balance in training programmes, low representation of women and no separate session on water and sanitation-related issues of women. Unfortunately, gender stereotyping reinforces patriarchal norms in handbooks and manuals of hygiene education. The strategy for gender mainstreaming in WASH includes conducting gender analysis, assessing gender-sensitive capacity-building needs, inclusion of gender right from the project planning stage and checking gender equity. Informal interaction and meeting with target groups, particularly women, semi-structured meeting with target groups to build their perspectives, and learning by doing exposure visits, demonstrations, campaigns, etc., can be effective tools to build the capacity. Thus, there is a need for a much more detailed analysis and structured review of training curriculum, and documenting best practices is needed to examine gender-related components in a better way. This can be undertaken by civil society organisations and development practitioners with the objective of policy and advocacy with the government and suggest ways to include gender in its entire capacity-building programme.

References

Ali, T., Sami, N. and Khawaja, A. 2007. Are unhygienic practices during the menstrual, partum and postpartum periods risk factors for secondary infertility? *J Health Popul Nutr*, 25: 189–194. Available at http://www.ncbi.nlm.nih.gov/pmc/articles/PMC2754005/ (accessed on 16 April 2013).

Bhardwaj, Sowmyaa and Archana Patkar. 2004. 'Menstrual Hygiene and Management in Developing Countries: Taking Stock', Junction Social, Mumbai. Available at www.washdoc.info/docsearch/title/163390 (accessed on 11 March 2013).

CCDU. 2007. *Information Guide to Village Panchayat Presidents on Drinking Water and Sanitation*. TNWSDB. Government of Tamil Nadu.

Fong, M.S., Wendy Wakeman and Anjana Bhushan.1996. 'Toolkit on Gender in Water and Sanitation; Gender toolkit series no.2', The World Bank, Washington D.C. Available at http://www-wds.worldbank.org/external/default/WDSContentServer/WDSP/IB/2001/01/20/000094946_00121301483084/Rendered/PDF/multi_page.pdf (accessed on 6 March 2013).

GOI. 2005. *Swajaldhara & TSC Capacity Building Programme for Indian States*. Centre for Good Governance and Uttaranchal Academy of Administration. Department of Drinking Water Supply, Ministry of Rural Development, Government of India.

———. 2010. *Rajiv Ghandhi National Drinking Water Mission: National Rural Drinking Water Programme, Movement Towards Ensuring People's Drinking Water Security in Rural India*. New Delhi: Department of Drinking Water Supply, Ministry of Rural Development, GoI. Available at http://rural.nic.in/sites/downloads/pura/National%20Rural%20Drinking%20Water%20Programme.pdf (accessed on 23 April 2013).

IFAD. 2007. *Gender and Water, Securing Water for Omproved Livelihoods: The Multiple Uses System Approach*. Rome: IFAD.

Indian Council of Medical Research. 2003. *Estimates of Maternal Mortality Rates in India and its States: A Pilot study, Ministry of Health and Family Welfare, GOI*, p. 1. Available at http://www.icmr.nic.in/final/Final%20Pilot%20Report.pdf (accessed on 15 September 2012).

Joshi, D. and Zwarteveen, M. 2012. Gender in drinking water and sanitation: an introduction, in M. Zwarteveen, S. Ahmed and S. RimalGautam (eds), *Diverting the flow: Gender equity and water in south Asia*. New Delhi: ZUBAAN.

Mahon, Thérèse and Maria Fernandes. 2010. Menstrual hygiene in South Asia: A neglected issue for WASH (water, sanitation and hygiene) programmes. *Gender & Development*, 18(1): 99–113.

MDWS. 2011a. Available at http://www.mdws.gov.in/aboutus (accessed on 11 March 2013).

———. 2011b. 'National Rural Drinking Water Mission'. Available at http://www.mdws.gov.in/sites/upload_files/ddws/files/pdf/RuralDrinkingWater_2ndApril_0_0.pdf (accessed on 11 March 2013).

———. 2011c. 'Nirmal Bharat Abhiyan'. Available at http://www.mdws.gov.in/sites/upload_files/ddws/files/pdf/Final%20Guidelines%20%28English%29.pdf (accessed on 11.March 2013).

———. 2011d. 'KRC Guidelines'. Available at http://www.mdws.gov.in/sites/upload_files/ddws/files/pdf/KRCguidelines_0.pdf (accessed on 11 March 2013).

———. 2011e. 'CCDU Guidelines'. Available at http://www.mdws.gov.in/sites/upload_files/ddws/files/pdf/CCDU.pdf (accessed on 11 March 2013).

———. 2011f. 'IEC Materials'. GOI. Available at http://www.mdws.gov.in/sites/upload_files/ddws/files/pdf/Pictoral%20guide%20on%20hygiene%20education%20%28hindi%29.pdf (accessed on 11 March 2013).

———. 2013. 'About NBA'. Available at http://tsc.gov.in/TSC/NBA/AboutNBA.aspx (accessed on 23 April 2013).

Millennium Project. 2006. 'Goals, Target and Indicators'. Available at http://www.unmillenniumproject.org/goals/gti.htm (accessed on 6 March 2013).

National Rural Drinking Water Programme (NRDWP). n.d. Available at http://www.mdws.gov.in/sites/upload_files/ddws/files/pdf/NRDWPintroduction.pdf (accessed on 23 April 2013).

Office of the Special Adviser on Gender Issues and Advancement of Women. 2006. *Gender, Water and Sanitation; Case Studies on Best Practices*. United Nations, New York.

PIB. 2005. 'Chapter 26; Rural Water Supply and Sanitation Programme'. Available at http://pib.nic.in/archieve/others/2005/nedocuments2005dec/ruraldevdec2005/Chapter26.pdf (accessed on 11 March 2013).

Planning Commission. 2011. Twelfth Five Year Plan—2012–2017; Report of the Working Group on Rural Domestic Water and Sanitation', GOI. Available at http://planningcommission.nic.in/aboutus/committee/wrkgrp12/wr/wg_indus_rural.pdf (accessed on 11 March 2013).

PRIA. 2002. Participatory Training: A Book of Readings. PRIA, New Delhi. Available at http://pria.org/publication/Participatory%20Training%20A%20book%20of%20readings%20A%20Resource%20book%20for%20front%20line%20workers%20of%20development%20organisations.pdf (accessed on 24 April 2013).

Ray, S. 2000, 'Training Manual on Gender Sensitization' Haryana Community Forest Project, Haryana Forest Department. Available at http://hcfp.gov.in/downloads/manuals/Training_Manual_on_Gender_Sensitization.pdf (accessed on 25 April 2013).

RGNDWM. 2004. 'TSC and Capacity Building: Activities and Module, Technical Note Series' MDWS. GOI. Available at http://www.mdws.gov.in/hindi/sites/upload_files/ddwshindi/files/pdf/TechnicalNoteonCapacityBuildingTSCSept%20 2004.pdf (accessed on 11 March 2013).

SaciWATERs. 2011. Report of the Fifth Gender, Water and Equity Training Workshop. SaciWATERs, Hyderabad, India

UNDESA. 2005. 'A Gender Perspective on Water Resources and Sanitation'. Background paper no. 2. Available at http://www.unwater.org/downloads/bground_2.pdf (accessed on 29 September 2012).

UNDP. 2006. Available at http://www.undppc.org.fj/_resources/article/files/19.pdf (accessed on 11 March 2013).

UNICEF. 2003a. 'Gender Mainstreaming in Water and Sanitation'. Available at http://www.unicef.org/wash/index_main_streaming.html (accessed on 7 March 2013).

———. 2003b. '10 key points to check for gender equity'. Available at http://www.unicef.org/wash/index_key_points.html (accessed on 11 March 2013).

———. 2012. 'Sanitation and Hygiene Advocacy and Communication Strategy Framework 2012-2017'. Available at http://www.mdws.gov.in/sites/upload_files/ddws/files/pdf/NSHAC_strategy_11-09-2012_Final.pdf (accessed on 20 September 2012).

UN Water, June 2006. 'Gender Water and Sanitation: A Policy Brief'. Available at http://www.unwater.org/downloads/unwpolbrief230606.pdf (accessed on 11 March 2013).

Victor, Berna Ignatius. 2006. *India: From Alienation to an Empowered Community—Applying a Gender Mainstreaming Approach to a Sanitation*. In United Nations (ed), *Gender, Water and Sanitation: Case Studies on Best Practices. Office of the Special Adviser on Gender Issues and Advancement of Women*. Department of Economic and Social Affairs, United Nations, USA, pp. 81–87. Available at at http://www.un.org/waterforlifedecade/pdf/un_gender_water_and_sanitation_case_studies_on_best_practices_2006.pdf (accessed on 20 September 2012).

Water Aid. n.d. 'Gender Aspects of Water and Sanitation'. Additional Resources. Available at http://www.wateraid.org/documents/plugin_documents/microsoft_word__gender_aspects.pdf (accessed on 6 October 2012).

WSP. 2010. 'Gender in Water and Sanitation'. World Bank. Available at http://www.wsp.org/sites/wsp.org/files/publications/WSP-gender-water-sanitation.pdf (accessed on 1 March 2013).

SECTION 2

Case Studies: Water

6

Gender Issues in Watershed Management

Suhas P. Wani, K.H. Anantha and T.K. Sreedevi

6.1 Introduction

The rain-fed areas in the semi-arid tropics are characterised by low and erratic rainfall, severe land degradation, low crop yields and high poverty. Watershed programmes are recognised as a potential engine for agricultural growth and sustainable development in rain-fed areas (Wani et al., 2003). The success and sustainability of watershed programmes are directly related to collective action and community participation (Wani et al., 2008; Sreedevi and Wani, 2007). Women are key players as managers and direct actors in managing natural resources in the watershed and addressing household food security and nutritional goals. However, too often, they play a passive role in decision-making processes because of their low educational levels, social customs and economic dependence. Though women share a major workload for managing the natural resources, the benefits of the watershed programmes largely bypass them, except where targeted income-generating and employment interventions have been undertaken (Sreedevi et al., 2009).

Increasing economic resilience of the poor is largely about enabling women to realise their socio-economic potential and to improve their life quality. Increasing women's participation in watershed management projects is critical to the long-term sustainability of development efforts. Since women rarely own or control productive assets, they are not looked upon as decision-makers in the management of natural resources (Seeley et al., 2000). Consequently, they do not receive their rightful compensation in terms of wages and ownership of productive assets and benefits accrued from them (Meinzen-Dick, 2004). In recent times, women self-help groups (SHGs), along with small savings, have unlocked a variety of avenues for income-generating activities (IGAs) for sustained income and livelihood opportunities (Anantha et al., 2009). These multifaceted avenues increase women's empowerment and system sustainability. Therefore, there is a need to sensitise policy makers and project implementing agencies on the core issues affecting women's participation in decision-making processes and the fair distribution of benefits. We use detailed case studies from Andhra Pradesh, India, to analyse what are critical factors essential for enhancing collective action for impact of watershed programmes, resulting in improving livelihoods and conserving natural resources.

6.2 Studying Watershed Villages

Three watersheds in Andhra Pradesh were selected, namely Adarsha watershed in Ranga Reddy district, Powerguda in Adilabad district and Janampet in Mahboobnagar district (Figure 6.1). The selection was based on the criteria that affect the performance of the watersheds, including social background, differential interventions for livelihood improvement, institutional arrangements and focus of the watershed. In these watersheds, the management was with community-based organisations (CBOs) and women had significant role to play. These watersheds were studied in detail for the process and the impact as well as the drivers of the success (D'Silva et al., 2004; Sreedevi et al., 2004; Wani et al., 2002, 2003). The details of the selected watersheds are described in Table 6.1.

In all the three cases, focus group discussions (FGDs) were organised. A common questionnaire was used but separately for women and men's groups. The FGD interviews revolved around issues related to women, particularly in terms of rights, workload, decision-making, access to information and earnings, social capital development, nature of the institutions, drivers of the success, and the type of benefits accrued, and their distribution amongst men and the women members. Information documented included collection, compilation and analyses to study the relationship amongst chosen variables (participation, workload, decision-making, access to information, social capital, institutions, etc.) and the type of interventions and approach adopted for watershed development.

In Adarsha watershed, Kothapally, an International Crops Research Institute for the Semi-Arid Tropics (ICRISAT)-led consortium adopted a farmer-centric, holistic, participatory approach for developing the watershed to increase the agricultural productivity and income. The meta-analysis of results and the interlocking constraints faced by farm

Figure 6.1:
Location map of selected study of watersheds in Andhra Pradesh, India

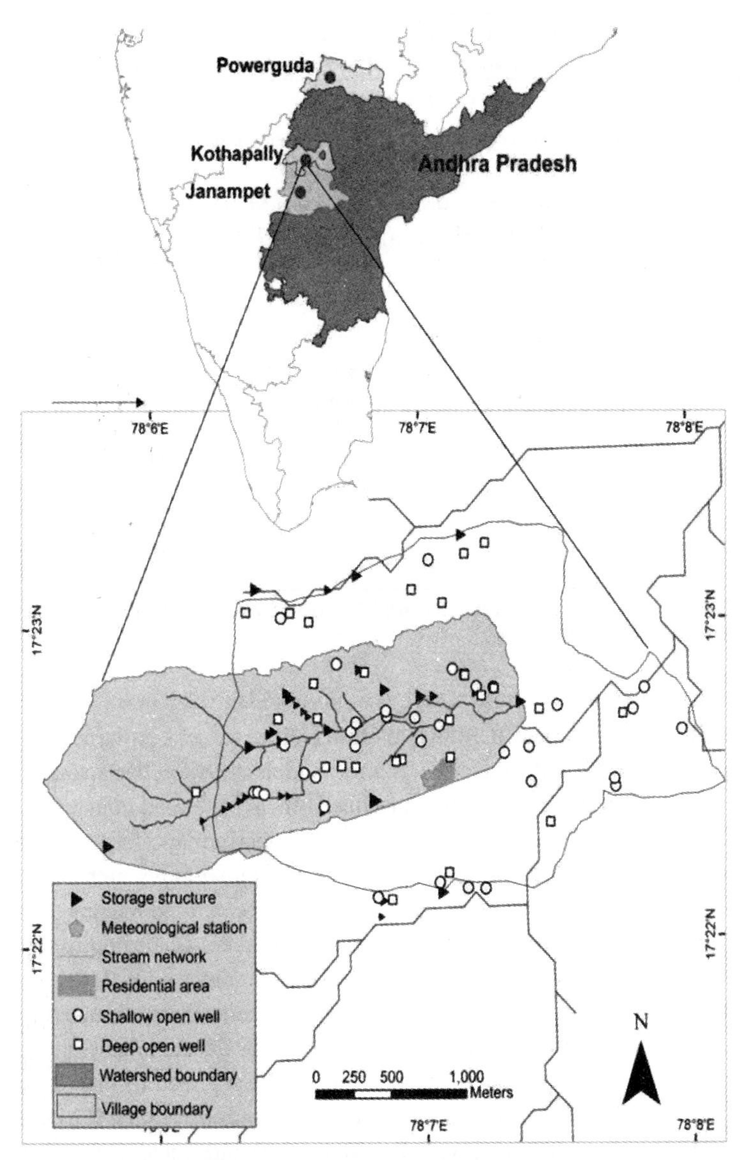

Source: Garg and Wani (2012).
Note: This map is not to scale and does not depict authentic boundaries.

Table 6.1:
Profile of selected watershed villages for the case study

	Adarsha Watershed, Kothapally	*Powerguda, Adilabad*	*Janampet, Mahboobnagar*
Proximity to city	Yes	No	Yes
Social background	Mixed community	Tribal—homogenous community	Mixed community
Watershed interventions	SWC + productivity enhancement + limited IGAs such as vermi-composting, nursery raising and livestock rearing	SWC + limited IGAs such as oil extraction unit, nursery	SWC + commercial activities—*Mahila Samakhya* undertake financing, highway restaurant, etc.
Managed by	Women SHGs for specific activities + WC representatives	Women SHGs, watershed implemented by women	SHGs are federated under *Mahila Samakhya* commercial activities
Emphasis	Productivity enhancement	Service provider using NRs and technologies	Commercial activities for income generation

IGAs, income-generating activities; SHGs, self-help groups; SWC, soil and water conservation.
Source: Sreedevi et al. (2009).

households prompted ICRISAT to launch strategic and on-farm development research. Meta-analysis is a methodology that collates research findings from previous studies and distils them for broad conclusions. There are several studies, which evaluated the performance of watershed programmes, and these studies were published either as research articles or reports. The watershed studies cover the entire rain-fed regions of the country and represent a wide range of environment according to their agro-ecological location, size, type, source of funding, rainfall, regional prosperity or backwardness and so on. The meta-analysis evaluated the impact of watershed programmes with the help of 626 micro-level studies to establish a higher degree of confidence in the analysed results for the comprehensive assessment of watershed programmes in India. ICRISAT-led community watershed work is based on the Integrated Genetic and Natural Resources Management (IGNRM) approach where activities are implemented at landscape level by the community (Wani et al., 2003).

Water management is used as an entry point to increase cropping intensity, increase productivity through enhanced water use efficiency, and also to rehabilitate degraded lands in the catchments, which helps in enhancing biodiversity, increasing incomes and improving livelihoods. Such an approach demands integrated and holistic solutions from seed to final produce. It involves various institutions and actors with divergent expertise varying from technical, social, financial, market, human resource development and so on.

In Powerguda, though the approach adopted was similar to the Adarsha watershed, it was distinct as the women SHGs implemented the watershed programme. Being a tribal area, the community had access to forest resources too. The Janampet watershed village is an advanced entity with the promotion of commercial activities for income generation. The SHGs at the village and *mandal* (block) levels have been federated and are known as the *Mahila Samakhya* Adarsha Women Welfare Society. This collectivisation has helped to increase women's bargaining power and branch out also into financial and political leverage. It also focuses on value-added services such as running a highway restaurant and other micro-enterprises.

6.3 Mainstreaming Gender Participation

ICRISAT introduced a new approach to watershed management that was followed in these watersheds. It incorporates gender-transformative activities based on a comprehensive, participatory and gender-sensitive analysis of vulnerability. This approach also helps to identify constraints with active participation from all stakeholders, especially women and vulnerable groups, to introduce need-based interventions to meet their desired goals. It recognises differential vulnerability within communities and households. The strategy builds on the existing knowledge and capacities of women members and is implemented with both men and women's participation. It also includes the most vulnerable groups in the community. Therefore, it promotes gender equality as a long-term goal while strengthening economic and social status as well.

Research and development (R&D) interventions at landscape level are conducted at these sites representing the different agro-ecoregions.

The entire process revolves around the four Es (empowerment, equity, efficiency and environment), which are addressed by adopting specific strategies prescribed by the four Cs [consortium, convergence, cooperation and capacity building (Wani et al., 2003)]. The consortium strategy brings together institutions from the scientific, non-government, government and farmers group for knowledge management and sharing. Convergence allows integration and negotiation of ideas among actors resulting in convergence of various programmes. These address the core issue of improving livelihood and protecting the natural resources (Figure 6.2). Co-operation encourages all stakeholders to harness the power of collective action. Capacity building enhances skills, leading to empowerment of communities and their sustainability.

Figure 6.2:
ICRISAT's strategy on women empowerment and livelihood adaptation through integrated watershed management

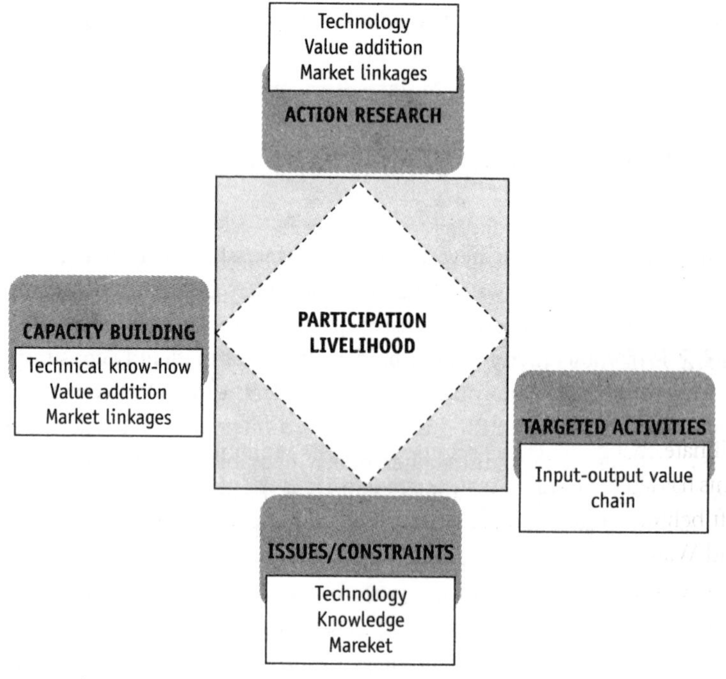

Source: Author's compilation.

6.3.1 Needs Assessment

Participatory watershed management is a multidisciplinary, multi-institutional approach for natural resource management and providing food security through diversification of livelihood options and increased productivity. Evaluation of a number of watershed programmes indicates the extent of peoples' participation and its importance in the success of development process and the role of institutions for enhanced community participation (Joshi et al., 2004, 2005, 2008). Watersheds with better community participation, including women, landless and vulnerable groups and sound technical inputs enhances the impact. Supporting policies are must for effective watershed development programmes (Joshi et al., 2009; Wani et al., 2008). Recognising the importance of participation, ICRISAT introduced need-based targeted interventions such as village seed banks, vermi-composting, nursery raising, dal making (dal preparation) to benefit women and vulnerable groups. ICRISAT believes in result-oriented action research for development. As a prelude to this objective, issues and constraints were identified for formulation of effective strategies within the community through FGDs, rapid rural appraisals, and formal and informal meetings. Since women make most household decisions, their understanding of the problem, needs and solutions were primary focal points of these discussions. This supports the twin objectives of participation and empowerment and achieves sustainability since they are designed to suit local conditions and skills of the beneficiaries.

6.3.2 Potential Opportunities

Climate change generates resource shortages and unreliable job markets. This has led to increased male migration. Thus, more and more women are left behind with additional agricultural and households duties (Sreedevi and Wani, 2007; Venkateswaran, 1992). Impoverished women's lack of access to and control over natural resources, technologies and credit mean that they have fewer resources to cope with seasonal and episodic weather and natural disasters. Consequently, traditional roles are reinforced. Girls' education suffers, and women's ability to diversify their livelihoods is diminished (Meinzen-Dick, 2004). Therefore, the new approach to watershed management, which focuses on productivity enhancement and livelihood improvement, provides the scope to identify potential opportunities for vulnerable and women stakeholders as per their needs and skill levels. This strategy helped large number of women beneficiaries to undertake several IGAs/strategies.

6.3.3 Income Alternatives

Building on social capital can make a huge difference in addressing rural poverty in watershed communities. In the watersheds we investigated, emphasis was laid on farm-based interventions such as preparation of vermi-compost, community seed banks and nursery raising as well as agriculture-related allied IGAs for landless and women's group members with the objective of increasing the income (Wani et al., 2003; Sreedevi et al., 2004). By adopting the principle of adding value to the produce to ensure that maximum proportion of market product price goes to the farmers and not to the middlemen, dal-making (dal preparation) proposition was also introduced in the Mentapally watershed of Andhra Pradesh. In this watershed, the SHG members have converted 60 kg of pigeon pea into dal and added ₹5,400 (US$108) worth value to their produce. Farmers worked at the charges to be paid to the SHG, which are lesser than the commercial mills and have recorded 90 per cent dal recovery. In addition to value addition, farmers have got the nutrient rich pigeon pea hulls to be used as animal feed (ICRISAT, 2004).

6.4 Gender Analysis of the Case Study Watersheds

6.4.1 Collective Action

The results revealed that the Integrated Watershed Management Programme (IWMP) approach adopted was different than the traditional watershed approach. In Adarsha watershed, Kothapally, and Powerguda, it was an integrated approach with emphasis on productivity enhancement of major crops (maize, pigeon pea and sorghum) and natural resource-related allied income enhancement activities.[1] In Powerguda, collective action was mainly for the service providing function. This was a step higher in the ladder of commercialisation over the Kothapally watershed where collective action was mainly for enhancing the productivity of their lands with a limited opportunity for direct economic gain. The nature and extent of collective action was also directly related with the awareness of the women members (Table 6.2). The women members in Janampet had a high level of awareness about running the highway restaurant. In the case of Powerguda, though the women leader was well aware, group

Table 6.2:
Gender impact analysis of three case studies in Andhra Pradesh, India

S. No.	*Description*	*Powerguda*	*Janampet*	*Kothapally*
1	Rights			
	Property	Men	Men/women	Men
	Financial resources of the family	Men	Women	Men
	Employment	Men/women	Men/women	Women
	Education	Men	Men	Men
	Social status of women	Medium	Good	Medium
	Awareness among women	Leader fully aware	Very good	Poor
	Agricultural decision-making	Men/women	Men/women	Men/women
	Resistance by men	Nil	Initial	Nil
2	Workload on women	Low	High	Medium
	Wages (Rs/day)			
	Men	50	50	50
	Women	30	30	30
	Load of invisible work	Same	Same	Same
	Workload on men	No	No	Yes
	Time spent on economic work by women	Low	High	Medium

(Table 6.2 Continued)

(Table 6.2 Continued)

S. No.	*Description*	*Powerguda*	*Janampet*	*Kothapally*
	Time spent on social/ community work	–	–	Medium
	Marketing of agriculture produce by women	–	Yes	–
3	Access to assets			
	Access to community assets	Men/women	Men/women	Men/women
	Access to credit	Women	Women	Women
	Access to income	–	Women	–
	Access to information	Yes	Yes	Yes
	Access to service	Nil	Yes	Yes
4	Control on financial resources	Low	High	Low
5	Self-confidence	Slowly building up	High	Low
6	Opportunities for exploration	Minimum	Very high	High
7	Understanding on health	Medium	High	Medium
8	Distressed migration	0	0	0
9	Driver identified	Leader	*Mahila Samakhya* (federation of women)	Improved water availability

Source: Adapted from Sreedevi et al. (2009).

members were not much aware about the operations as well as the rules and procedures to be adopted. In Janampet, the approach for improving livelihoods was on the commercial scale, and direct economic gain was the main purpose. The women SHGs were federated and the collective action were at a macro-level. This reaped the benefits of common learning, exposure and opportunity to interact with more and diverse group members as well as reduced transaction costs. In Kothapally and Powerguda, the collective action was restricted at small group levels in the village, and exposure for the members was restricted and transaction costs were higher in terms of load on the leadership.

6.4.2 Women Rights and Gender Equity

The impact of the model/approach adopted was distinctively evident in the case study villages (Table 6.2). In terms of rights, it was revealed that Janampet ranked on the top for property rights. Here, women held the property rights along with men. In Kothapally and Powerguda, the property rights were with the men except in the exceptional cases where women headed households due to the death of a male member. The nature and the extent of collective action provided different exposures for the members. In Janampet, the commercial nature of collective activities resulted in women's control of family financial resources. In Kothapally as well as in Powerguda, although women family members earned the money, the control of family financial resources rested with men. In Kothapally, the women group activities provided employment to women members mainly because of the type of activity undertaken. In Powerguda and Janampet, the collective action of women created employment opportunities for women as well as men.

6.4.3 Education and Social Status

In the investigated watersheds, the right to education rested more with men. Efforts to tilt education in favour of women will need a longer time. In Kothapally, the education of boys and girls is similar and no child labour exists in this village. Every school-age child is in school. However, in Powerguda and Janampet, child labour exists. In Powerguda, indigenous women are now aware about educating their daughters. The literacy rate among the population is 45 per cent. Interestingly, female literacy (52 per cent) is higher than male literacy (48 per cent). About 35 children from 3 to 6 years go to *Anganwadi* (preparatory school). Eighteen girl children out of total 31 children were studying in a boarding school some 50 km away from their hamlet. Among children, aged 6–20 years, the literacy percentage is 62 without much gender differentiation. Some of the families are sending their children to English medium schools, paying at least ₹20 per month. There is at least one member from each household attending school. This shows awareness of the value of education in the village. The social status of women in all of the three study watersheds was better than the normal watershed village.[2] However, amongst the three watersheds, Janampet women enjoyed higher social status than the women in Kothapally and Powerguda.

6.4.4 Women Workload and Wages

In terms of workload on women, it was higher in Janampet than in Kothapally and Powerguda. Looking at the extent of commercial activities undertaken by the women SHGs, the workload was higher in Janampet. Although Powerguda SHGs undertook a higher scale of commercial activity than the Kothapally SHGs, the workload on Kothapally's women was more than in Powerguda. The Powerguda women employed men for undertaking specific activities and paid higher wages for men as compared to women considering the nature of the work undertaken. Similarly in Janampet, women members compensated their family labour in the field by hiring additional labourers from the market. This financial independence permitted women SHGs to work out alternate arrangements to reduce their workload. However, in all the three watersheds the wage differences between men and women labourers existed, with men being paid higher (₹50 per day) than the women labourers (₹30 per day). Traditionally, men and women undertake specific farm activities and as observed in Powerguda, women felt that the specific jobs done by men need to be paid differently. In Janampet, only women undertook marketing of agricultural produce whereas in Powerguda and Kothapally men took up this activity (Table 6.2).

6.4.5 Women's Empowerment and Decision-making

In all the three watersheds only women SHGs had access to financial credit, as per the current policy of the government. The women members had good access to information. However, new opportunities for exploration were directly in line with the extent of commercial nature of the activities undertaken. In all the three case studies, the new watershed approach encompassing productivity enhancement and livelihoods approach had direct and positive impact on reducing the distressed migration of men and women from the villages. For example, in all the case study villages, before the implementation of watershed activities, men and women alike were migrating to nearby urban places in search of their livelihood especially after the rainy season. However, this situation is reversed with the implementation of need-based interventions that are providing income-earning opportunities in all the seasons. In case of Kothapally, the awareness amongst the members was low, as most of the banking and financial transactions had to be done at a *mandal* (block) level bank situated 15 km away from the village. Men and women took decisions related to agriculture jointly. This is a step in the right direction for sustainable management of natural resources. Men did not resist the progressive measures of women in all these case study watersheds although there was some resistance by the male family members in Janampet initially.

6.5 Results

Reducing rural poverty in watershed communities is evident in the transformation of their economies. In Adarsha watershed, Kothapally, the ICRISAT model ensured improved productivity with the adoption of cost-efficient water harvesting structures as an entry point for improving livelihoods. Crop intensification with high-value crops and diversification of farming systems are leading examples that have allowed households to achieve production of basic staples and surplus for modest incomes. Enhanced participation of the vulnerable groups like women and the landless through capacity building and networking was observed. Kothapally women who were actively involved in the initiative through SHGs, initially focusing on vermi-composting, have now gone on to finance a diverse range of small-scale enterprises ranging from tree nurseries to tailoring. Women SHG members became master trainers to neighbouring villagers on preparation of vermi-compost and travelled to other states as resource persons. The SHGs are common in the watershed villages and provide income and opportunities for empowerment of women. In 2001,

Figure 6.3:
Income stability and resilience effects during drought year (2002) in Adarsha watershed, Kothapally, Andhra Pradesh, India

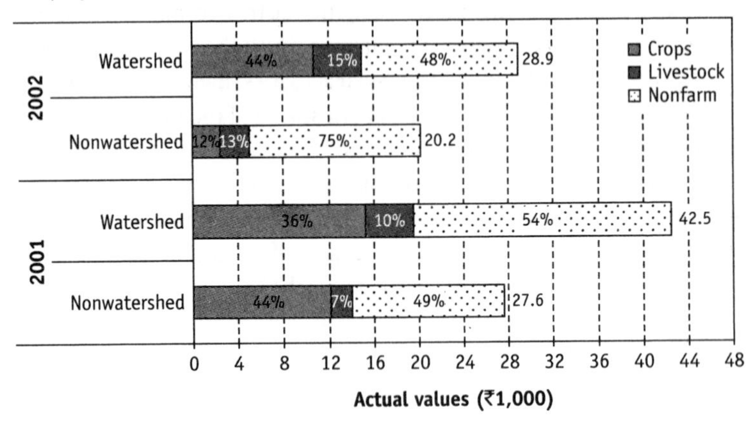

Source: Wani et al., (2008).

the average village income from agriculture, livestock and non-farming sources was ₹42,500 compared with the neighbouring non-watershed village with ₹27,600 (Figure 6.3). Due to additional groundwater recharge, about 200 ha in rainy season and about 100 ha in post-rainy season were cultivated with different crops and cropping sequences. The productivity of maize increased 2 to 2.5 times under sole maize and four-fold under maize-pigeon pea intercropping system. The area under maize-pigeon pea and maize-chickpea has increased more than three-fold and two-fold, respectively. Farmers could gain about ₹16,510 and ₹19,460 from these two systems, respectively. The average household net income has increased to ₹15,400 within the watershed area as compared to ₹12,700 outside the watershed area (Wani et al., 2003).

The SHGs with the watershed programmes in Powerguda, Andhra Pradesh, had six-fold higher savings than those without such programmes in Adilabad district. The introduction of improved land management practices such as BBF and bullock-drawn tropicultor (a multi-purpose wheeled tool carrier), along with high-yielding cultivars increased agricultural productivity by 20 times to 350 per cent. Powerguda farmers, particularly many women, learned new techniques in planting, land preparation and intercropping. Many of them grew vegetables for the first time. Over 3 years, there was a remarkable change in cropping patterns, shifting from cotton to soybean and vegetables (D'Silva et al., 2004). A woman SHG managed an oil extracting machine[3] to support IGAs in the community. Seeds of Pongamia, neem and other trees are crushed in this machine to extract oil that is sold in the market. The oil mill has become an important source of income to Powerguda. The women SHG planted about 8,500 Pongamia trees in 2002 and 2003 and 10,000 in 2004 to augment the oilseed supply in future. Since October 2003, Powerguda has discovered a new IGA in tree nurseries. The community decided to invest in a Pongamia nursery ₹30,000 received from the World Bank as part of environmental service payment. For the first time, 147 tons of carbon dioxide was sold from India to the World Bank (D'Silva et al., 2004).

An average family income increased by 77 per cent in 3 years from ₹15,677 in 1999–2000 (before the government invested in watershed development) to ₹27,820 in 2002–2003. Seasonal migration from villages has ended totally, or is negligible. It appears that watershed and agricultural development, complemented by other investments, have provided sufficient employment and income opportunities for the rural people to escape poverty and to stay in the village (D'Silva et al., 2004).

Since 1999, Powerguda has charted a new path of development using watershed management as the growth engine, women SHGs as the institutional anchor, and a total ban on the consumption of alcohol in the village. These steps have enabled Powerguda to march ahead of the other neighbouring hamlets. The people, especially the women leaders, are very proud that they have been able to outperform other villages in social, financial, institutional and environmental development. Powerguda is distinguished from other hamlets due to the strong leadership provided by women through SHGs. Three of the four SHGs are run by women who dominate most of the development activities in the village. Trust, social cohesion, a sound local leadership and democratic functioning of local institutions are among the features of social capital in Powerguda.

Powerguda was unique in that the women SHGs were the dominant institutions in the village. These SHGs had gone farther than thrift. They delivered some of the services, which previously were the responsibility of government agencies. SHGs in the village run a Pongamia nursery with a capacity for 20,000 saplings. SHGs also replaced private contractors in implementing some of the public works. For example, local residents under the management of SHGs built all the watershed structures in the village. These activities had helped to build the confidence of the SHG leadership while also increasing the coffers of the group. In the watershed contracts, there was an opportunity to save between 18 and 25 per cent of the cost of the structures.

The Janampet watershed village is a step further than the Powerguda and Adarsha watersheds. With the supporting policies from the government, the SHGs at the village, *mandal* (block) and district levels are federated to increase their bargaining power as also financial and political leverage. The women SHG federation provides a forum for women to discuss common issues. The SHG members consider the unity and solidarity among women to be one of the most important benefits of SHG membership. At the *mandal*-level (block) federation meetings, women of different castes and class come together. This solidarity enables them to share their problems and seek help. Also by standing guarantees for SHGs, the federations can help the SHGs to borrow money from financial institutions at lower interest rates. These loans are particularly useful for value-added services such as running a highway restaurant and other micro-enterprises. The federation takes care of bookkeeping and training functions of SHGs. The *Mahila Samakhya* Adarsha Women Welfare Society is a federation of women SHGs and Janampet SHGs are members of the federation. The impact in terms of increasing the family incomes, building the social capital as well as trust amongst the women members from Janampet is superior to the Powerguda or Adarsha watersheds.

6.6 Drivers of Success

The drivers of success varied in all the case study watersheds:

- In Powerguda, the success was directly associated with the strong and capable leadership provided by Ms. Subhadrabai. It may be noted that through training and exposure, Subhadrabai could become a very capable leader though she did not have formal schooling. She was able to channel the energies of fellow women for the sustainable development of the village using natural resources.

- In Kothapally, the main driver of the growth and success increased due to the availability of water resources resulting in increased agricultural productivity and triggered agriculture-related allied activities such as vermi-composting. The women groups in the watershed are collaborating together and progressing to achieve the sustainability through more collective action and exploring new opportunities to increase the income from the collective action.
- In Janampet, it was the collective action and supporting government policy, which enabled the women SHGs to undertake commercial activities successfully with the help of the leadership. Janampet watershed witnessed the highest level of community participation where collective action or collegiate mode of participation is reached. This level of participation in the collective action is quite sustainable and the group can overcome most of the problems through their collective wisdom and opportunities.
- Looking at the process of community participation, the mode of participation starts or is initiated through a co-opting or contractual process and slowly moves towards cooperative, consultative, collaborative and finally reaching to the successful collective action (Sreedevi et al., 2009). It was found that Janampet watershed was on the highest ladder of community participation, whereas Powerguda watershed on the lower ladder.

6.7 Conclusion

The preceding discussion revealed that the integrated watershed management approach promotes gender equality. It is clear that the mere presence of women in watershed committees is not enough for achieving women's welfare. New approaches such as productivity enhancement in community watersheds and integrated IGAs along with specific targeted activities such as availability of water and energy sources are needed to reduce drudgery. These case studies reveal that unless targeted income-generating and employment interventions for women, landless and the vulnerable groups are introduced, the economic resilience of the poor and meaningful participation may not be possible. Therefore, promotion of need-based IGAs such as micro-enterprises and value-addition activities are essential to achieve socio-economic potential and improve the quality of life. Higher the commercialisation of IGAs, better the women's status and decision-making power in the families and villages. For harnessing gender power, holistic livelihood approach in the community watershed programmes is needed rather than traditional compartmental approach of rainwater harvesting and conservation. To address the issue of gender inclusiveness, watershed programmes should look beyond land development activities. It should take into account the diverse ways in which rural people make their livelihoods from both agrarian and non-agrarian-based IGAs. There is a need to make available the technical know-how and make women aware of such technologies through regular skill upgrading trainings. Enhanced awareness of women's rights through deliberate efforts is critical for sustainable development of watersheds by harnessing the power of women equitably. Considering the basic rule of collective action that under stress, people cooperate better and greed for higher personal benefits affects the collective action, there is a need to harness the gender power through harmony in the watersheds at all the levels, starting from the family to watershed.

The lessons learned from the new productivity enhancement and livelihood improvement approach are the guiding force in formulating new common guidelines (Government of India, 2008) for watershed management in the country. Based on the approach developed by ICRISAT and its partners, the government of Andhra Pradesh and Karnataka have adopted a similar approach in bridging the yield gaps and enhancing household income. Further, the Government of India has established 13 model watersheds, by adopting new common guidelines, across nine states in India as sites of learning for which ICRISAT provides technical backstopping. The new common watershed guidelines provide resources and policy support to address equity issues of gender and vulnerable groups. However, without concrete actions such as promotion of need-based IGAs, holistic livelihood approach in the community watershed programmes, regular skill upgrading trainings, etc., by the implementing and coordinating agencies, these provisions would not mean much. There is also a need for creating awareness among all the partners involved in the projects, such as government and non-government implementing agencies, CBOs, SHGs and women groups. This can be enhanced by regular trainings and increasing the number of exposure visits by all agencies involved.

Acknowledgements

The authors would like to thank the Asian Development Bank, Government of India and Government of Andhra Pradesh along with other consortium partners for their support and participation in carrying out the present study in participatory watershed management.

Notes

1. Under productivity enhancement, farmer-based soil and water conservation measures implemented in individual fields were broad-bed and furrow (BBF) landform and contour planting to conserve *in situ* soil and water; use of tropicultor for planting, fertiliser application and weeding operations; field bunding; and planting Gliricidia on field bunds to strengthen bunds, conserve rainwater and supply nitrogen-rich organic matter for *in situ* application to crops (Wani et al., 2003).
2. Normal watershed refers to those implemented by following erstwhile watershed guidelines without focusing either on productivity enhancement or on livelihood activities.
3. Worth ₹375,000 provided by Integrated Tribal Development Agency (ITDA).

References

Anantha, K.H., Wani, S.P. and Sreedevi, T.K. 2009. Agriculture and allied micro-enterprise for livelihood opportunities, in Wani, S.P., Venkateswarlu, B., Sahrawat, K.L., Rao, K.V. and Ramakrishna, Y.S. (eds), *Best-bet Options for Integrated Watershed Management. Proceedings of the Comprehensive Assessment of Watershed Programs in India*, p. 312. Patancheru 502 324, Andhra Pradesh, India: ICRISAT.

D'Silva, E., Wani, S.P. and Nagnath, B. 2004. The making of new Powerguda: Community empowerment and new technologies transform a problem village in Andhra Pradesh. Global Theme on Agroecosystems Report No. 11, p. 28. Patancheru 502 324, Andhra Pradesh, India: International Crops Research Institute for the Semi-Arid Tropics.

Garg, K.K. and Wani, S.P. 2012. *Opportunities to Build Groundwater Resilience in the Semi-Arid Tropics*. Groundwater DOI: 10.1111/j.1745-6584.2012.01007.x

Government of India. 2008. *Common Guidelines for Watershed Development Projects*. p. 60. New Delhi: Department of Land Resources, Ministry of Rural Development, Government of India.

International Crops Research Institute for the Semi-Arid Tropics (ICRISAT). 2004. Improved Livelihood Opportunities through watersheds, Completion Report, April 2002 to June 2004. p. 80. New Delhi: Andhra Pradesh Rural Livelihood Project (APRLP) and Department for International Development.

Joshi, P.K., Jha, A.K., Wani, S.P., Joshi, Laxmi and Shiyani, R.L. 2005. *Meta-analysis to assess impact of watershed program and people's participation*. Research Report 8. Comprehensive Assessment of Watershed Management in Agriculture. Patancheru, India: International Crops Research Institute for the Semi-Arid Tropics; and Manila, Philippines: Asian Development Bank.

Joshi, P.K., Jha, A.K., Wani, S.P. and Sreedevi, T.K. 2009. Scaling-out community watershed management for multiple benefits in rainfed areas, in S.P. Wani, J. Rockström, and T. Oweis (eds), *Rainfed Agriculture: Unlocking the Potential. Comprehensive Assessment of Water Management in Agriculture Series*, pp. 276–291. Wallingford, UK: CAB International.

Joshi, P.K., Jha, A.K., Wani, S.P., Sreedevi, T.K. and Shaheen, F.A. 2008. Impact of watershed programs and conditions for success: A meta-analysis approach. Report No. 46 Global Theme on Agroecosystems, 24 pp. Andhra Pradesh, India: International Crops Research Institute for the Semi-Arid Tropics.

Joshi, P.K., Vasudha Pangare, Shiferaw, B., Wani, S.P, Bouma, J. and Scott, C. 2004. Socioeconomic and policy research on watershed management in India–Synthesis of past experiences and needs for future research. Global Theme on Agroecosystems, Report No. 7. Patancheru, Andhra Pradesh, India: International Crops Research Institute for the Semi-Arid Tropics.

Meinzen-Dick, Ruth, DiGregorio, Monica and McCarthy, Nancy. 2004. Methods for studying collective action in rural development. *Agricultural Systems* 82: 97–214.

Seeley, J., Batra, Meenakshi and Sarin, Madhu. 2000. *Women's participation in watershed development in India*. Gatekeeper Series no. 92, 20 pp. London, US: International Institute for Environment and Development.

Sreedevi, T.K., Shiferaw, B. and Wani, S.P. 2004. Adarsha watershed in Kothapally: understanding the drivers of higher impact. Global Theme on Agroecosystems Report no. 10, 24 pp. Patancheru 502 324, Andhra Pradesh, India: International Crops Research Institute for the Semi-Arid Tropics.

Sreedevi, T.K. and Wani, S.P. 2007. Leveraging institutions for enhanced collective action in community watersheds through harnessing gender power for sustainable development, in S. Mudrakartha (ed), *Empowering the Poor in the Era of Knowledge Economy*, pp. 27–39. Ahmedabad, Gujarat, India: VIKSAT.

Sreedevi, T.K., Wani, S.P. and Nageswara Rao, V. 2009. Empowerment of women for equitable participation in watershed management for improved livelihoods and sustainable development: An analytical study, in Amita Shah, Wani, S.P. and Sreedevi, T.K. (eds), *Impact of Watershed Management on Women and Vulnerable Groups*. Proceedings of the Workshop on Comprehensive Assessment of Watershed Programs in India, 25 July 2007. Andhra Pradesh, India: International Crops Research Institute for the Semi-Arid Tropics.

Venkateswaran, S. 1992. *Living on the Edge: Women, Environment and Development*. New Delhi: Friedrich Ebert Stiftung.

Wani, S.P., Joshi, P.K., Raju, K.V., Sreedevi, T.K., Wilson, J.M., Shah Amita, Diwakar, P.G., Palanisami, K., Marimuthu, S., Jha, A.K., Ramakrishna, Y.S., Meenakshi Sundaram, S.S. and Marcella, D'Souza. 2008. *Community Watershed as a Growth Engine for Development of Dryland Areas. A Comprehensive Assessment of Watershed Programs in India*. Global Theme on Agroecosystems Report No. 47, p. 156. Patancheru 502 324, Andhra Pradesh, India: International Crops Research Institute for the Semi-Arid Tropics.

Wani, S.P., Pathak, P., Tam, H.M., Ramakrishna, A., Singh, P. and Sreedevi, T.K. 2002. Integrated watershed management for minimizing land degradation and sustaining productivity in Asia, in Zafar Adeel (ed), *Integrated Land Management in Dry Areas*. Proceedings of a Joint UNU-CAS International Workshop, 8–13 September 2001, (Beijing, China). pp. 207–230.

Wani, S.P., Ramakrishna, Y.S., Sreedevi, T.K., Long, T.D., Wangkahart, Thawilkal, Shiferaw, B., Pathak, P. and Kesava Rao, A.V.R. 2006. Issues, Concepts, Approaches and Practices in the Integrated Watershed Management: Experience and lessons from Asia in Integrated Management of Watershed for Agricultural Diversification and Sustainable Livelihoods in Eastern and Central Africa: Lessons and Experiences from Semi-Arid South Asia. Proceedings of the International Workshop held 6–7 December 2004 at Nairobi, Kenya. pp. 17–36.

Wani, S.P., Singh, H.P., Sreedevi, T.K., Pathak, P., Rego, T.J., Shiferaw, B. and Shailaja Rama Iyer. 2003. Farmer-Participatory Integrated Watershed Management: Adarsha Watershed, Kothapally India, An Innovative and Upscalable Approach. A Case Study, in R.R. Harwood and A.H. Kassam (eds), *Research Towards Integrated Natural Resources Management: Examples of Research Problems, Approaches and Partnerships in Action in the CGIAR*, pp. 123–147. Washington, DC, USA: Interim Science Council, Consultative Group on International Agricultural Research.

7

Gender and Governance: A Case of Jalswarajya Project

Aditya Bastola

7.1 Introduction

The immediate period following the 1990s in India saw reforms in all the sectors to streamline development economics. The government's efforts to provide safe drinking water was criticised, as problems of drinking water were intensified along with the scarcity of water resources (Kulkarni et al., 2008). The lack of access to safe drinking water for the deprived and lack of people's participation in water programmes brought a drastic shift in governmental policies in India.

At a global level, this was clear in The New Delhi Declaration (1990) that laid the principles for institutional reform and participation of women at all institutional levels. Similarly, The Dublin Statement (1992) recognised the role of women in water management and considered water as a finite resource. It urged that people should recognise water as an economic good. The principles of Dublin Statement were again reiterated at the UN Earth Summit, Rio (1992).

By this period, the World Bank in partnership with the Government of India implemented three large drinking water projects[1] in India (James, 2004). Before any structural change, the World Bank rationalised sector reform in provisioning of drinking water to address poverty, via decentralised community-based initiatives (World Bank, 1999, 2001). This assessment of the drinking water project indicated the need for a fundamental shift from the traditional supply-driven approach to demand-driven approach and showed that the poor were willing to pay for water services (Joshi, 2004).

When the Government of India made the 73rd and 74th Constitutional Amendments in 1992, the PRIs was seen as a platform for decentralised drinking water resource management at the community level. The VWSC was established by the drinking water project and this transferred the responsibility onto the villagers to operate and maintain the water infrastructures. In the drinking water sector, the model of decentralised water governance was first implemented by Swajal Project (in Uttar Pradesh, now Uttarakhand) and got replicated through the Sector Reform Pilot Project (SRPP). The SRPP was implemented in 1999 across 67 districts in 26 states of India (Cullet, 2009a, 2009b, 2011; James, 2004; Joshi, 2004; WSP-SA, 2002).

The SRPP clearly laid the foundation of a national demand-driven approach, as it involved 10 per cent community contribution to the capital cost and 100 per cent Operation & Maintenance (O&M) costs towards the users. The project guidelines made it clear to state governments implementing the sector reform drinking water project to raise their own financial capabilities. This was primarily because the Constitution of India purviews rural drinking water supply as a state subject with the caveat of ensuring the right to adequate potable water under the purview of right to life (Panickar, 2007).

Consequently, the State Government of Maharashtra (GoM) was the first state in India to implement sector reform project on a large scale. Initially, the GoM implemented Aaple Pani Project[2] through the support of the German Government Development Bank (KfW) in three districts; later the model was replicated as the Jalswarajya Project, with the support of World Bank, across the other 26 districts of the state.

Jalswarajya Project (meaning 'water independence') was considered the largest sector reform project in terms of its outreach, the strategy to draw women's participation and the establishment of decentralised institutions. The decentralised institutions were VWSC, Women Development Committee (WDC) and Social Audit Committee (SAC) in each village to effectively manage water supply services and to empower women. In fact, Jalswarajya Project was unique in its position, with the establishment of three decentralised institutions, it promoted 50 per cent women's participation within VWSC and SAC and 75 per cent in the WDC (GoM, 2003).

Women participating in the Jalswarajya Project are viewed as moving beyond the private spheres and being key players in community decision-making processes making them actors of good water governance. How effective is their participation to address gender strategic needs or even to voice their concerns within the framework of good water governance processes? This chapter seeks to answer these questions in light of empirical evidence from the villages where Jalswarajya Project was implemented.

Theoretically, rural women under the Jalswarajya Project formed a majority not just within VWSC, but also extending to other institutions, which were more political in nature. This constituted a landmark move beyond the constitutional amendments (73rd and 74th Amendments in 1992 and 1993, respectively) of one-third representation of women in local governments. What made this project unique was the non-negotiable principle of 50 per cent women's representation and one-third representation of the marginalised caste groups in the decentralised water institutions. The rationale behind this provision was that women have a central role in the provisioning of drinking water to meet household needs, and are not involved within the decision-making processes of development project.

The chapter is divided into six sections. The first section outlines the policy shifts in drinking water sector, followed by the second section that discusses the decentralised drinking water project from a gender perspective. The third section outlines the methodology employed in the study. The framework in good water governance is outlined in the fourth section. The background of Jalswarajya Project and its institutional mechanisms are presented in the fifth section, and its implication at the field level is presented empirically as the sixth section. The final section presents a discussion and conclusion to the issue raised in which implications of the present study is highlighted.

7.2 Decentralised Drinking Water Project and Gender

The efforts so far undertaken to promote decentralised drinking water projects at community level, has to a large extent promoted concepts of good water governance so that gender interests (strategic and practical) (Molyneux, 1985) are addressed in an equitable manner. The voices of the poor and the marginalised are challenged through the establishment of institutions at the local level (Cleaver and Hamada, 2010).

After the adaptation of demand-driven approach, there was a dramatic shift towards the promotion of women's participation within the decentralised water institutions. Although, traditionally, the drinking sector had always seen women carrying head-loads of water, walking long distances to fetch water and performing the domestic work (Ahmed, 2005; van Wijk-Sijbesma, 1998), in the post water sector reform women were seen as central water managers and water an economic good. This shift in rural drinking water supply projects marked women's involvement to be associated with their domestic role to supply drinking water at household level (Cleaver, 1998a). The water users handled 100 per cent O&M costs of the water supply infrastructures. This shows that the whole process of institutionalising women's participation or collective action within the drinking water project has mainly been to justify women's saved time in collecting water and then this time could be allocated for increased opportunity to instigate income generation activities (IGAs) (Cleaver, 1998b; World Bank, 1993).

Inclusion of women within the decentralised water institutions has been to justify the project norms so that project rules are followed, without considering whether women's participation within the water management activities are meaningful to bring change in their conditions (Joshi, 2004; Kulkarni et al., 2008). As rightly marked by Cleaver and Hamada (2010), this is also related with the orientation of water policies that is mainly aimed at service delivery and ignoring the social factors such as the caste, class, ethnicity, religion and gender.

It is evident from the Swajal Project that when it was implemented it promoted 33 per cent of women's participation within the VWSC, yet women's participation remained low. This was because the project did not specifically address social and power relations in their design, due to which women's participation in the project did not significantly affect the project outcomes (Prokopy, 2005: 1816). Singh (2006) states that women's participation in drinking water projects is largely driven by two factors, more so in India and South Asia. The first is their individual factor that is related with their disinterest in politics, low educational attainments, lack of confidence, lack of exposure, lack of leadership skills, and physical and health conditions. Although these individual factors play a significant role, the other social factors that impose restrictions (typically in homogeneous/matrilineal communities) on individual behaviour cannot be ignored as they govern the process for effective outcomes.

While the individual factor tends to be guided by norms, values and laws, women's participation within the decentralised water institutions are driven by social or institutional factors. These include patriarchal values and norms, caste-based discriminations, religion, lack of support from husband, dominance from in-laws or the domestic responsibilities that exclude women to attend or being vocal in public meetings.

In the SRPP, women within the decentralised institutions were incorporated from the *Panchayat* board members, while influential husbands nominated the others. Women had no choice in becoming the members of the decentralised institutions, nor the technology of drinking water supply; the male members in the village controlled the decision-making processes (Swayan Skishan Prayog, 2002).

Interestingly, in rural drinking water projects, women's participation and the time saved in collecting water have been incentivised for IGAs (Cleaver, 1998b). Therefore, most women saw this incentive as an opportunity to achieve economic benefits and therefore they participated in the drinking water project. The collectivisation of women was limited to just being members of the self-help groups (Datar, 2008) rather than articulating their concerns in water committee meetings (Ranadive-Deshmukh, 2005).

The women and the poor who already had little control of the decision-making saw this opportunity for enhancement of their status. Challenging the rural elites' hold was difficult. However, it is important to note, that when democratic decentralisation is coupled with policies of privatisation and pricing of the resources, the poor and the marginalised may find it difficult to negotiate and contest the power of rural elites (Johnson, 2003). This is because the poor and the marginalised have to face with different layers of social hierarchies such as caste, class and gender that impinge one's access and control over common resources especially over water resource.

7.3 Methodology

The findings of this chapter draw from the field work conducted in 2008. The field area included 12 villages across four districts of Maharashtra. The objective of the research was to examine women's participation within the decentralised water institutions with respect to its impacts at household level. The villages in the survey method were selected on the basis of Jalswarajya Project being implemented as Phase I, Batch I and those villages which had completed the project implementation were termed 'exit' by the Reform Sector Project Management Unit (RSPMU), Maharashtra, which was responsible for monitoring the progress of the project at the state level.

A mix of quantitative and qualitative methodologies was applied in the survey method. The primary data was gathered through administered questionnaire (as quantitative data) and the qualitative data was gathered through improvised Participatory Rural Appraisal (PRA) techniques and the Focus Group Discussion (FGD). One FGD was conducted per village. Women who were members of the decentralised institutions comprised the 248 respondents of this study. In the FGDs, a mixed group of women from the village participated to understand the gender dynamics that evolved due to the project.

7.4 The Gendered Framework on Good Water Governance

Water governance process is viewed as a tool not only to address the shortcomings of the broader governance process (such as accountability, participation and among others) but also to meet the Millennium Development Goals (MDGs) (Cleaver and Franks, 2008; Cleaver and Hamada, 2010). Although governance involves from people's participation to utilising the available resources, it does not operate in a vacuum and can be achieved through constant negotiation (Cleaver and Franks, 2008). When the notions of good governance were promoted within the water sector, accountability with participation, transparency and responsiveness (Rogers and Hall, 2003) were recognised as important criteria of governance by the UN (Cleaver and Franks, 2008). The rationale of having such a strategic choice is due to problems in resource allocation and lacunae in expenditure-tracking mechanism, monitoring and evaluation, and people's participation (World Bank, 2004). The promotion of good governance was a part of the local self-government that evolved in the 1990s (Cornwall, 2010). However, within governance processes there are several aspects such as interpersonal relationships that can draw, deny or limit participation and be very political in tone. Governance as a process involves change in the behaviour of individuals, which are often mediated through values, norms and laws (Rogers and Hall, 2003). Therefore, the water governance framework drawn from Rogers and Hall (2003) by Cleaver and Franks (2008) states that, it is largely held by factors as given in the following subsections.

7.4.1 Resources and Mechanisms

Resources include material and non-material resources, where people exercise power, determine inequalities and draw rules for allocation.

Although mechanism involves institutions, it also extends to the relationship amongst the stakeholders. In the context of the study, the mechanism covers the ways of organising access to water, more related with establishment of institutions, its rules, and promotion of SHGs and among others. This could include rules and the norms as the principles, technological choices and water tariffs among others.

The Jalswarajya Project promotes a uniform technological choice (pipeline water supply services through single or multi-village schemes with a minimum litres per person per day norm). The process through which the practice of good water governance is brought about in Jalswarajya Project is related with the project's non-negotiable principles. These include the mandatory 50 per cent women participation within the VWSC and SAC and 75 per cent in the WDC, women taking leaderships as treasurer of the VWSC and the SHGs involved in the collection of the water tariffs, and handling the O&M of the water supply infrastructures. These non-negotiable principles in the project have generated rules and norms within the water institutions for effective delivery of water services and bring change in women's subordinate conditions.

7.4.2 Outcomes

To a large extent, the outcomes of governance process are strongly held by other factors (like social factors). It implies that women groups as part of SHGs mobilise themselves to collect the water tariffs, which usually was conducted by the village *Panchayat*, or women take the role of water mechanic, a domain that was previously perceived as a man's role. The change in women's role facilitated by the project can bring about significant change in the lives of women.

The changing social role extends the notions of good water governance into outcomes where women can voice their needs, have the capacity to collectivise and make an impact on their livelihood options (Cleaver and Hamada, 2010).

7.4.3 Processes

In this water governance process, there are several changes that take place within the three components of governance given earlier; it could be a conscious change or unconscious but eventually it brings change in its outcomes, which may or may not be participatory. As Cleaver and Franks (2008) state, the change could result in increased access to existing resources or access to a completely new set of resources, followed by an establishment of a new mechanism. The negotiation that takes place within decision-making and action brings changes in resources, mechanisms and outcomes of water governance.

In the context, it is the negotiation that women as individuals and as groups take to bring change in their conditions, thereby addressing gender strategic and practical interests.

7.4.4 Actors or Agents

There is no governance if there are no interactions amongst groups that are associated in service delivery. It is these actors that provide a leveraging component in the water governance that evolve out of these four components that constantly interact. These agents in the form of women, men, community members, government officials, NGOs get shaped or shape resources, mechanisms and outcomes through a range of gender-specific processes (Cleaver and Franks, 2008). Therefore, the framework has been applied to understand how Jalswarajya fits into the processes of good water governance and to examine the implications for women.

7.4.5 Shaping Good Governance Practices within Jalswarajya Project

Through the recent move for decentralisation within the drinking water sector, new institutions have been created at the village level; particularly in the Jalswarajya Project, three institutions were formed (VWSC, WDC and SAC). A minimum of 50 per cent women's participation within the decentralised committee meetings is mandatory within the project. The VWSC has the financial authority over the project capital, so it strategically appoints women as treasurer of the committee to handle the project funds, thereby seeking more women's participation.

To follow a good governance model, Jalswarajya Project has designed different layers of partnership building. The local community decides whom to contract for construction of pipeline and select the technical support persons (TSPs) and Non-Government Organisation (NGOs) to carry out the overall implementation of the scheme. The local contractor is answerable to the VWSC members who in turn are answerable to the community members. The process of governance ranges from delivery of public service from the government to those local private contractors, rule settings from the government to the local community members.

In this process, both women and men within the decentralised committees are entrusted with their respective roles and responsibilities from the mobilisation of community members for the 10 per cent contribution, planning, designing, budgeting, implementing to monitoring and evaluation. Specifically, the VWSC are supposed to mobilise villagers for community action, collect the capital contribution from the project beneficiaries, participate in capacity-building trainings, submit reports to the Village *Panchayat* and execute contracts and procure materials required for construction of water and sanitation infrastructures.

The SAC is responsible for continuously observing the functioning of the VWSC and its sub-committees, so that project rules/principles such as inclusion, equity, cost-effectiveness, transparency among others are not violated. The members of WDC are responsible for preparing women development plans, carrying out repair work of the water infrastructure and collecting water tax from the project beneficiaries. Women's SHGs undertake the responsibility of collecting the water tax from each project beneficiary. Hence, members from all these three committees are answerable to the villages (GoM, 2003).

These are resources and mechanism within the framework of good water governance, and the next section addresses how it operates.

7.5 Implication of Jalswarajya Outcomes for Women

The Jalswarajya Project rules (as non-negotiable principle) state that a minimum of six women are to be members of the SAC and VWSC. In one of the villages, when women members of the three decentralised institutions were selected as respondents, it was found that only four women were members of the SAC that was listed on the *Panchayat* building wall. When asked for the reason as to why the rules of the project were unmet, the president of the VWSC stated, 'more women could be nominated to satisfy the minimal criteria by simply writing two names of the women from the village.'

The statement of the VWSC president revealed the deeply entrenched hegemony of power vis-à-vis the newly created representative processes. This was clearly illustrated when almost three-quarters of the women from the water institutions stated their membership was largely through the recommendation of the Village *Sarpanch* (village head) or the *Gram Sevak* (village-level worker). Nonetheless, about one-fourth of the women interviewed during the field investigation were surprised to learn that they were members of the Jalswarajya committee.

In reality, the rural elites motivated the women including their family members to participate so that they could benefit from the project activities. Women from lower caste groups were mobilised to participate in the project. These women formed the representatives for caste and women's groups in the decentralised institutions. The findings from the field (Table 7.1) revealed that three-quarters of the women members belonged to SCs, STs, OBCs and the *Vimukta Jati* (de-notified tribes) and nomadic tribes (VJNT), and only one-quarter were members of the general caste groups.

In the project, the reservation policy outlined (GoM, 2003: 54) that not less than 30 per cent of the seats were to be reserved for SC, ST, OBC

Table 7.1:
Distribution of women's caste in decentralised water institutions

Caste	*Frequency*	*Percentage*
General	63	25.4
Schedule caste	29	11.7
Schedule tribe	84	33.9
Other backward classes	51	20.6
Vimukta Jati and Nomadic tribes	21	8.4
Total	248	100.0

and VJNT representation in the VWSC. The data clearly indicates that, there is higher representation of backward caste groups within the decentralised institutions. Since five of the villages were classified as tribal by RSPMU, the non-tribal villages were selected to further probe the actual caste representation. The VSWC as a significant committee[3] of Jalswarajya Project was selected to analyse the caste representation of women from the non-tribal villages.

Table 7.2 shows that of a total of 42 members, 25 are representatives from the SC, ST and OBC categories though the actual representation as per reservation (30 per cent) of the SCs, STs and OBCs equates to 13 members. The data clearly indicates that the backward community members

Table 7.2:
Non-tribal village-wise representation of women's caste category in VWSC

		Caste Classification in VWSC					
Districts	*Villages*	*General*	*SC*	*ST*	*OBC*	*VJNT*	*Total*
Buldhana	Sagoda	2	0	3	1	0	6
	Wadi	1	2	1	2	0	6
Nashik	Kikhware Kh.	3	1	1	1	0	6
Osmanabad	Ansurda	2	0	0	3	1	6
	Bedarwadi	2	0	0	1	3	6
	Bhonja	3	2	0	0	1	6
	Aasu	4	1	0	1	0	6
Total		17	6	5	9	5	42

were merged within the women category, as there was no increase within the backward men category.

This shows women belonging to the SCs, STs and OBCs were merged as women representatives to fulfil both the 50 per cent reservation of women and 30 per cent reservation of marginalised caste groups. Since there was no government regulation to quantify women's representation from the reserved caste groups other than the overall 30 per cent reservation, women from the marginalised caste groups were brought to participate as proxies for the marginalised men. In doing so, the 50 per cent reservation of women was used by the rural elites to represent women as caste groups. This reduced the overall women's representation, as 30 per cent of SCs, STs and OBCs belonged to the women's representation.

The decentralised water governance, which aimed for efficiency through people's participation, was likely to be using reservation policy as taxonomy to capture resources and the benefits meant for the development of the poor by the upper/dominant caste and class groups in Jalswarajya Project. Nonetheless, the rural elites from the VWSC controlled not only the decision to appoint the TSP, and the NGOs, but also the dissemination of the project account details.

If we look into the process/mechanisms how these institutions operate, it would be interesting to reflect on women's participation and the committee meetings held pre- and post-implementation of the project. When the Jalswarajya Project was implemented, the committee meetings were organised regularly (on a monthly basis) and the *Mahila Gram Sabhas* were held prior to the *Village Gram Sabha*. But after the project phased out, there were no water committee meetings. A woman recalling this moment states, 'women are again neglected in the Gram Sabha, as it was in the earlier times.' Although the Jalswarajya Project rules/non-negotiable principles were addressed, women's participation within the decentralised institutions was largely seen to justify the project norms. Consequently, women coming together attending meetings and sharing issues was short-lived.

When the women participated within the decentralised institutions, their representation outnumbered the men, and a majority of these women were members of the SHG. The statistical association between the variables, Jalswarajya strategy and participation in committee meeting was statistically significant (significant at 5 per cent level, $P = .007$) (Bastola, 2012: 105).

The promotion of SHGs for women empowerment has differing views (Baltiwala and Dhanraj, 2008; Jakimow and Kilby, 2006; Ranadive-Deshmukh, 2005). In Jalswarajya villages, the SHGs were promoted before the implementation. So when the project was implemented, some women saw their participation to help to bring change in women's conditions. Interestingly, considering the Indian social context that determines women's access to drinking water (Ahmed, 2005; Kulkarni et al., 2008), caste as an independent variable was statistically examined with women's participation to bring change in women's condition. The results were statistically significant (at 5 per cent level, $P = .000$) (Bastola, 2012: 107).

To sum up, in women's mobilisation, the SHG was a catalyst (Jakimow and Kilby, 2006; Ranadive-Deshmukh, 2005). Therefore, a dummy variable (as becoming a member of the SHG) to control any unaccounted effect inherent to participation was statistically tested. The results in Table 7.3 show that the original bivariate association between 'caste' and 'women's

Table 7.3:
Effect of SHG membership on caste and participation to bring change in women's conditions

			To Bring Change in Women's Condition		
Member of the SHG			*No*	*Yes*	*Total*
Yes	Caste	General	24	23	47
		SC	9	10	19
		ST	34	20	54
		OBC	8	37	45
		VJNT	10	5	15
	Total		85	95	180
$\chi^2 = 23.576$		df = 4	$P = 0.000$		$C = 0.340$
No	Caste	General	10	6	16
		SC	8	2	10
		ST	26	4	30
		OBC	3	3	6
		VJNT	6	0	6
	Total		53	15	68
$\chi^2 = 7.995$		df = 4	$P = 0.092$		$C = 0.324$

χ^2 = Chi square value; df = degree of freedom; P = significance; C = contingency coefficiency participation to bring change in women's condition was unchanged by the third variable 'member of the SHG' in its first categories that was 'Yes'.

This shows that opinions about bringing change in women's conditions are unchanged if they are members of SHG, but conditional if these women are non-members. Although, SHGs have a significant role, caste identity to some extent cannot be ignored as women from the lower caste/class groups, who suffer from the patriarchal values and norms see Jalswarajya Project as a mechanism to emancipate.

Therefore, when women participated in the committee meetings to bring change in their conditions, their participation was largely related to satisfy the project rules. A majority of the women were unaware about the project budget, as most were not interested to know the costs incurred. The responsibility for collection of water tariffs were not handled by the women group, instead the *Gram Panchayat* operated and maintained the water supply infrastructure.

Women had begun to gain confidence to participate in Jalswarajya Project, through the promotion of SHGs, and had gained confidence to speak on issues, but most of women's issues related with addressing water needs. Therefore, women's participation within the project mainly related to fulfil their gendered roles and responsibility that was socially ascribed such as the reproductive role. Nevertheless, a woman in the village reported, 'Today men have started to accept women coming in the front and taking initiatives until the power structures or some important structure is not hindered by the women.' This means addressing women's practical needs are accepted by the men folk, as it does not challenge the patriarchal power structures. Therefore, unless gender relations are not prioritised in water governance, the implication of bringing women as SHG or decentralised institution members in Jalswarajya Project for women empowerment can have minimal impact on good water governance and gendered outcomes.

The framework developed from Cleaver and Franks (2008) has been modified as per the resources, mechanisms, outcome, process and actors present in Jalswarajya Project and is presented in Figure 7.1.

In fact, through the whole governance process it was felt that women attending meetings and undertaking the cleanliness activities at the village level, women's household roles were extended to community work. A minimal change was observed in terms of gendered outcomes to determine their livelihood options through economic benefits, as the men or the in-laws largely made decisions at household level.

Figure 7.1:
Water governance in Jalswarajya Project

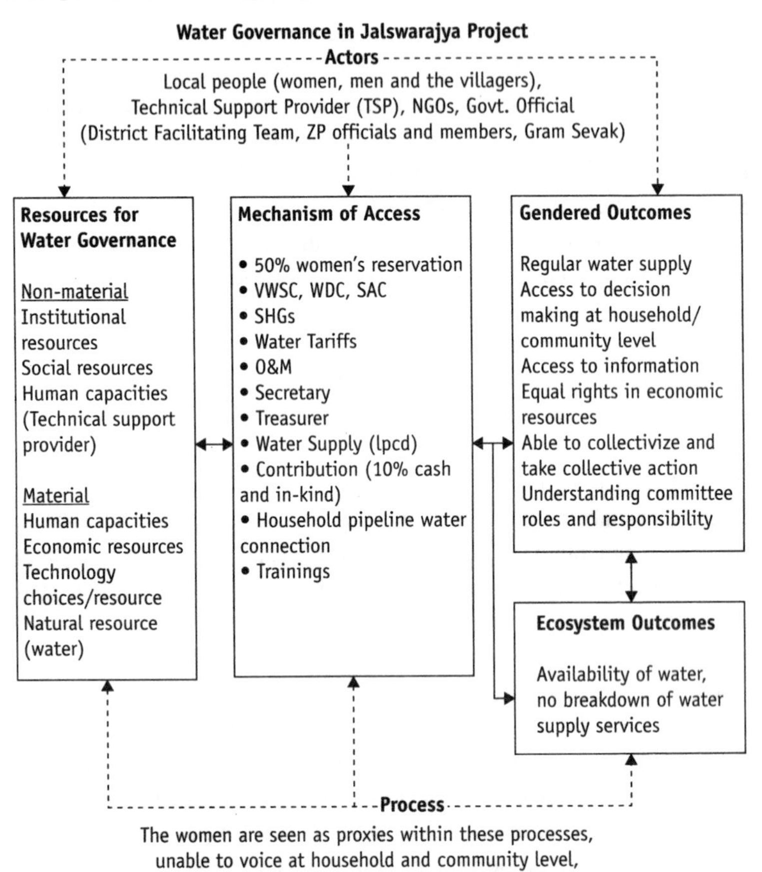

Source: Cleaver and Franks (2008), modified p. 164.

7.6 Conclusion

The paradigm shift in water governance was to disguise the realignment of the state's neo-liberal roles. The outcome of such processes shifted the burden onto the poor and the women to pay for water services. It also allowed the socio-economic and political elites to shrink their responsibilities to the so-called weaker sections, and to continue to exclude those they had already socially marginalised.

The resources as part of the incentives in drinking water project especially with the time-saved phenomena (Cleaver, 1998b) had generated women's participation in the hope of bringing changes in their socio-economic conditions. But women's participation within the institutions was seldom realised because much of these institutions were often embedded with local power and social structures (Goetz, 1997).

The rural elites have found this process as an opportunity to capture power and control over the project funds and use women from the marginalised caste groups as another example of proxy representatives of the PRIs within water governance. Thus, over-emphasising of women's participation without addressing caste and class factors for gender outcomes actually masked the power differentials and political interest within the water institutes.

The resources provided by the project were seen as an opportunity to escape from their sufferings by the lower caste groups. With that hope, women participated as SHG members, but most of the gender practical needs were prioritised over the strategic needs. As a result, the outcomes of governance processes aimed at empowerment did little at household and community levels to challenge the patriarchal values and norms.

Although other mechanisms such as SHGs created a momentum for women to voice their practical needs, there is a need to continue the efforts of capacity building through linking SHGs, and the community mobilisation process with other government schemes and programmes. The purpose is Jalswarajya Project should not be a one-time activity as social and cultural change is not a time-specific process. If not, activities such as SHGs would be used by the rural elites as a medium to serve other purposes such as to mobilise women to participate in the decentralised institutions or invite SHG members in decentralised water committee meetings.

Further, to prevent proxy representation of women within water institutes, it is essential that appropriate gender representation be in place even within the one-third caste representation. This will provide the poor women and men equal opportunities to participate, thereby minimising the scope of power capture by the rural elites.

In nutshell, to ensure sustainability of similar projects, there is a need to put emphasis on women's participation within class and caste factors to ensure gender practical and strategic needs. Strengthening the mechanisms for appropriate gender representation will lie down the checks and balances during different phases of project implementation. Sustenance of women's collective efforts and their capability to influence decisions at village and household levels is important for good water governance processes.

Notes

1. Maharashtra Rural Water Supply and Environmental Sanitation Project (1991–1999); Karnataka Integrated Rural Water Supply and Environmental Sanitation Project (1993–2000); and The Swajal Project (1996–2002).
2. Aaple Pani Project was implemented in 1999–2000 in three districts of Maharashtra (Pune, Aurangabad and Ahmednagar) to increase rural household's access to improved and sustainable water and sanitation services and to institutionalise decentralisation of rural water supply and sanitation service delivery to rural local governments and communities.
3. The VWSC is responsible for procurement, execution and supervision of the water and sanitation project. The committee is entrusted with the responsibility to coordinate with the support organisations and the overall operation and management of the water and sanitation infrastructures.

References

Ahmed, Sara. 2005. Why is gender equity a concern for water management? in Ahmed, S. (ed), *Flowing Upstream: Empowering Women through Water Management Initiatives in India*, pp. 1–50. New Delhi: Centre for Environment Education.

Bastola, A. 2012. Review of Jalswarajya Project through Gender and Development Perspective, Unpublished Ph.D. Thesis submitted to University of Pune, Pune.

Batliwala, Srilatha and Dhanraj, Deepa. 2008. 'Gender myths that instrumentalize women: A view from the Indian front line, in Cornwall, A., Harrison, E. and Whitehead, A. (eds), *Feminism in Development: Contradictions, Contestations & Challenges*, pp. 21–34. New Delhi: Zubaan.

Cleaver, F. 1998a. Choice, complexity, and change: Gendered livelihoods and the management of water. *Agriculture and Human Values*, 15(4): 293–299.

———. 1998b. Gendered incentives and institutions: Women, men and the management of water. *Journal of Agriculture and Human Values*, 15(4): 347–360.

Cleaver, F. and Franks, T. 2008. Distilling or diluting? Negotiating the water research—policy interface. *Water Alternatives*, 1(1): 157–176.
Cleaver, F. and Hamada, K. 2010. "Good" water governance and gender equity: A troubled relationship. *Gender and Development*, 18(1): 27–41.
Cornwall, Andrea. 2010. Introductory overview—buzzwords and fuzzwords: Deconstructing development discourse, in Cornwall, A. and Eade, D. (eds), *Deconstructing Development Discourse: Buzzwords and Fuzzwords*, pp. 1–18. Oxford: Oxfam GB.
Cullet, P. 2009a. *Water Law, Poverty, and Development: Water Sector Reforms in India.* New Delhi: Oxford University Press.
———. 2009b. New policy framework for rural drinking water supply: Swajaldhara guidelines. *Economic and Political Weekly*, 44(55): 47–54.
———. 2011. Evolving regulatory framework for rural drinking water: Need for further reforms, in Infrastructure Development Finance Company (IDFC) (ed), *India Infrastructure Report 2011: Water: Policy and Performance for Sustainable Development.* New Delhi: Oxford University Press.
Datar, C. 2008. Decentralised governance, gender & affirmative action in rural drinking water management. *Indian Journal of Industrial Relations*, 44(2): 209–226.
Goetz, A.M. 1997. *Getting Institutions Rights for Women in Development.* London: Zed Books.
Government of Maharashtra (GoM). 2003. *Jalswarajya, Maharashtra Rural Water Supply and Sanitation Project—Project Implementation Plan.* Mumbai: Water Supply and Sanitation Department, September.
Jakimow, T. and Kilby, P. 2006. Empowering women: A critique of the blueprint for self-help groups in India. *Indian Journal of Gender Studies*, 13(3): 375–400.
James, J.A. 2004. *India's Sector Reform Projects and Swajaldhara Programme: A Case of Scaling up Community Managed Water Supply*, IRC International Water and Sanitation Centre, the Netherlands. Available at http://www.irc.nl/content/download/19658/242147/file/IndiaJamesFINAL_AJJ_ed_2.pdf (accessed on 1 March 2013).
Johnson, C. 2003. Decentralisation in India: Poverty, Politics and Panchayati Raj, ODI Working Paper 199, UK: Overseas Development Institute (ODI).
Joshi, D. 2004. *Secure Water—Whither Poverty? Livelihoods in the Demand Responsive Approaches (DRA): A Case Study of the Water Supply Programme in India.* London: Overseas Development Institute, November. Available online at http://www.odi.org.uk/resources/docs/3850.pdf (accessed on 12 May 2012).
Kulkarni, S., Ahmed, S., Datar, C., Bhat, S., Mathur, Y. and Makhwana, D. 2008. Water Rights as Women's Rights? Assessing the Scope for Women's Empowerment through Decentralised Water Governance in Maharashtra and Gujarat, SOPPECOM, Utthan and TISS, Supported by International Development Research Centre, Canada.
Molyneux, M. 1985, Mobilisation without emancipation? Women's interest, the state and revolution in Nicaragua. *Feminist Studies*, 11(2): 227–254.

Panickar, M. 2007. State Responsibility in the Drinking Water Sector—An Overview of the Indian Scenario, Working Paper IELRC 2007–06. Geneva: International Environmental Law Research Centre (IELRC).

Prokopy, S.L. 2005. The relationship between participation and project outcomes: Evidence from rural water supply projects in India. *World Development*, 33(11): 1801–1819.

Ranadive-Deshmukh, J. 2005. *Women in Self Help Groups and Panchayati Raj Institutions: Suggesting Synergistic Linkages*. New Delhi: Centre for Women's Development Studies.

Rogers, P. and Hall, W.A. 2003. Effective Water Governance, TEC Background Papers No. 7, Global Water Partnership Technical Committee (TEC). Stockholm: Global Water Partnership.

Singh, N. 2006. Women's participation in local water governance: Understanding institutional contradictions. *Gender, Technology and Development*, 10(1): 61–76.

Swayam Shikshan Prayog. 2002. *A Rapid Appraisal of Sector Reforms in the RWSS—Rural Water Supply and Sanitation Programme in Maharashtra*. Mumbai (unpublished Mimeo).

The Dublin Statement on Water and Sustainable Development. 1992. Issued Following the International Conference on Water and Environment (ICWE), Dublin, Ireland, January 31 1992. Available online at http://www.un-documents.net/h2o-dub.htm (accessed on 12 May 2012).

The New Delhi Statement. 1990. *New Delhi Statement, Global Consultation on Safe Water and Sanitation*. New Delhi, October 1990. Available online at http://www.ielrc.org/content/e9005.pdf (accessed on 12 May 2012).

van Wijk-Sijbesman, C. 1998. Gender in Water Resource Management, Water Supply and Sanitation: Roles and Realities Revisited, Technical Paper Series No. 33. Delft, the Netherlands: IRC International Water and Sanitation Centre.

Water and Sanitation Program—South Asia (WSP-SA). 2002. Implementing Sector Reforms: A review of selected state experience, *Jal Manthan: a rural think tank*, 6 June 2002, Water and Sanitation Program—South Asia, New Delhi.

World Bank. 1993. *Water Resources Management*. Washington D.C.: The World Bank.

———. 1999. *Rural Water Supply and Sanitation*, World Bank South Asia Region Rural Development Sector Unit in Collaboration with the Government of India, Ministry of Rural Areas and Employment (The Rajiv Gandhi National Drinking Water Mission), and DANIDA. New Delhi: Allied Publishers.

———. 2001. *World Development Report 2000/2001: Attacking Poverty*. New York: Oxford University Press.

———. 2004. *World Development Report 2004: Making Services Work for Poor People*. Washington D.C.: The World Bank.

8

Unleashing the Gender Differentials in Water Management: The Rural Milieu

Pradeep K. Mehta and Niti Saxena

8.1 Introduction

Water is a basic necessity. It is one of the most important components of survival. However, water scarcity is a rampant problem across the globe. As Nash (2010) explains, 18 per cent of the world's population lacks access to safe drinking water. According to the UN estimation, by the year 2025, around two-thirds of the world's population will face water shortage (Aureli and Brelet, 2004). Even though, there have been several initiatives to address the problem, not much change has happened. The Millennium Development Goals aim to halve the number of people without sustainable access to safe drinking water and basic sanitation by the year 2015 (DESA, 2010). However, this seems to be falling far behind due to the manifold rise in population and the expected growth in the coming years. As per FAO, over 230 million people live in 26 countries classified as water-deficient (UNEP, 2012). Water scarcity brings with it numerous socio-economic implications. Largely, women and girls are entrusted with the task of collecting and managing water for domestic household tasks such as bathing, cleaning and cooking. In rural contexts, men usually require water for irrigation and livestock rearing. In the urban context, these aspects are diluted. Resultantly, women are the forbearers of water management. Thus, whenever there is scarcity of clean water, the poor and women are the first ones to bear the consequences (UNEP, 2012). A report by FAO (2000) states that despite the participation of men in agriculture, women actually look after the farming of irrigated crops. For instance, it estimates that women farmers grow half of the world food. In developing countries where access to entitlements is a far cry for many, women and girls walk an average of 6 km a day to fetch water. With around 20 kg of weight on their heads, they trudge barefoot to fend for the water needs of the family (IAEA, 2007). This scenario is re-lived every day in the rural parts of India, especially in the regions where water scarcity is rampant. Case studies from various states in India such as Rajasthan highlight the drudgery women undergo in fetching water, especially in the face of water scarcity (NCW, 2005).

Mewat is one such region where this is a common sight. A district within the prosperous state of Haryana in northern India, its geographical locale and the local culture revert it to centuries of backwardness. A semi-arid region located at the foothills of the *Aravalis*,[1] the region struggles with ground water salinity, which is pervasive. Largely inhabited by Meo Muslims,[2] the region rates extremely low on literacy levels (56.10 per cent) and high on fertility, with the average family size being 7.5 members per household (Census, 2011). A break-up of the literacy figures reveals a grimmer picture with female literacy at 37 per cent. While there are several reasons attributed to girls being out of school such as parents not considering education to be important for them, their engagement in taking care of siblings or doing other household tasks, including fetching water, which is one of the major reasons due to which girls drop out of school. Water scarcity in the region has other psychological and economic dimensions attached to it, which, potentially impact every household. Being a patriarchal society, the onus of decision-making regarding the choice of water source lies with the men. The extensive participation of women in water-related tasks are diluted to an extent that it finds no place in discourses related to the planning, construction and maintenance of water sources in the region. Resultantly, the drudgery that women undergo to manage water for domestic chores is massive (Mehta et al., 2011). The implications are long drawn and reflect on their health and economic status.

Against this backdrop, the chapter addresses several questions: What are the gender roles in management and usage of water resources? How do perspectives differ on choice of water sources, usage and maintenance? How do differing water situations impact the time and effort put into procuring water for domestic and commercial tasks? How do men and women perceive water work differently in terms of their emphasis to quantity and quality of water? What are the gender differentials with respect to paying for the maintenance and usage of water sources?

The chapter is divided into three sections. The first section provides a backdrop of the study and the methodology used to capture primary data from the field. The second section discusses results. This section is further subdivided into three parts. The first subsection outlines the usage of water sources and related impact assessed on the lines of gender differentials. The gendered roles in decision-making on water-related issues are explicated in the second subsection followed by evaluating the socio-economic impact of differing water situations in the third subsection. The next section summarises the key findings of the study followed by recommendations.

8.1.1 Methodology

To capture perspectives on planning, construction and maintenance of water sources, water management responsibilities both for domestic and commercial purposes, gender disaggregated data was collected from both male and female respondents. With differing water situations existing in the region, data has been collected from both potable and saline water villages. The rationale for the purposive selection is to understand the differences in water management between villages with differing water situations, the extent of drudgery undergone by the inhabitants, and their coping strategies. Responses have been obtained from 180 respondents comprising 90 males and 90 females selected randomly from three potable and three saline water villages. To obtain a holistic perspective, qualitative tools have also been employed. These include problem tree discussions, chapatti making and focus group discussions with both men and women. Discussions with village water and sanitation committees were not possible as these committees are absent in the sampled villages.

8.2 Results and Discussion

The results on water facilities in the villages with differing water situations and their related impact on the inhabiting population are discussed below. Gender-specific roles with respect to decisions related to water issues are explored and the eventual impact discussed.

8.2.1 Usage of Water Sources and Related Impact

Differing water situations impact coping strategies in water management and handling. Often, the assumption is that in villages with potable ground water, the drudgery undergone in procuring water for domestic purposes would be less as compared to saline water villages. This is because in the latter, the reliance on ground water is nil and therefore, procurement of water is largely through piped supplies from other regions. Location of water sources in the sampled population however, indicates the contrary. In potable water villages, the water source for 7 per cent households is outside the homestead. In saline water villages, the responses are scattered with more than half having water sources located inside home. Discussions reveal that due to the absence of ground water sources and piped supplies in saline villages, inhabitants have underground water tanks in their compounds, which are filled through purchased water tankers. The fact is reaffirmed with the distribution of water sources in the villages. Figure 8.1 clearly indicates that in saline water villages, purchased water is the major source for drinking water and other domestic consumption. Contrastingly, in potable water villages, the sources are scattered with government water

Figure 8.1:
Distribution of water sources

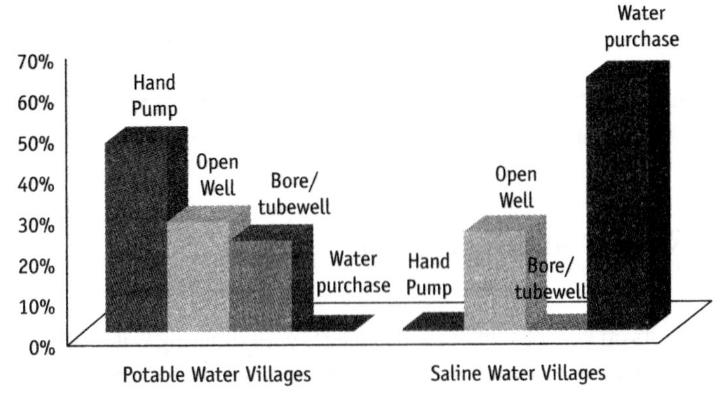

Source: Primary Data (2012).

supply. These include open well, hand pumps and tube/borewells. The high dependency on purchasing water has financial implications. The average amount spent by a household in a year on purchasing water amounts to ₹6,476 (US$119.22[3]) and ranges from 10 to 18 per cent of the total annual expenditure of the household in the region. For the poor households with limited income, purchasing water for survival places a lot of burden. As is evident from Figure 8.1, in potable water villages, borewells and tube wells are also used as sources for procuring water for all purposes. The data reflect that installation of a tube well or a borewell amounts to around ₹30,000–50,000 (US$552.30–920.50). Thus, it becomes clear that only affluent households can afford the installation. Resultantly, the poor households in potable water villages are dependent on the rich to share water with them. Sharing of water is a practice observed in both types of villages. Discussions reveal that as per the local values and norms, giving water to the poor for free is considered as a good deed, which will fetch them blessings of god.

With respect to the maintenance of community water sources in both potable and saline water villages, several households attribute this to be the responsibility of the concerned government department or the *Sarpanch*, that is, the village headman. Differentiation is observed in responses obtained from both the villages. While in saline water villages, majority of respondents mentioned that it should be the government or the *Sarpanch* or *Chowkidar* (watchman), in potable water villages, 47 per cent households mention that the villagers should maintain community water sources as they are the eventual users of the facility. Difference in the level of ownership towards maintenance is seemingly because in most of the saline water villages, the community water sources have been lying as redundant entities. Water supply is erratic and therefore, water doesn't suffice for all households. Hence, the lack of ownership is starkly visible in saline water villages. While talking about the willingness to pay for the maintenance of community water sources, a similar trend is observed. In potable water villages, 88 per cent of households are willing to contribute in cash or kind to repair a damaged community source. In saline water villages, however, this willingness reduces to 33 per cent households agreeing to contribute for repair of community water sources. The observed trend may be related to the argument presented earlier.

Lack of water in the villages is reflected in the average water requirement of every household with respect to drinking and other domestic purposes. While in saline villages, the requirement is 37 litres of water for drinking; in potable water villages, this requirement goes up to 47 litres. One of the plausible reasons is that access to more quantity of water eases out the amount of water available at one's disposal and hence the incremental water quantity. In the case of saline villages, water requirements have to be contracted due to the limited amount of access and the cost implications associated with water procurement. Furthermore, the average water requirements for other purposes range from 177 litres in potable villages to 192 litres in saline villages. The magnitude of water requirements for other purposes, however, reflects the grave implications on households that do not purchase water and have to walk long distances to fetch water.

Distance of a water source from home is directly proportional to the time spent in procuring water as well as the effort that one has to put into its procurement. This implies that more the distance of water source from the place of living, more is drudgery of procurement. While there is no formulated indicator to measure drudgery, the time and effort put into procuring water can be directly related to the labour undertaken.

Even though, the number of water sources in saline water villages is limited, the time spent in fetching water in a day is comparatively less in the saline water villages (Table 8.1). While in saline water villages, it takes around 148 minutes to fetch water; in potable water villages, the time spent is almost 1.60 times, that is, 237 minutes. This is related to the

Table 8.1:
Time spent in water-related activities[4]

	Potable Water Villages (Approximate)	*Saline Water Villages (Approximate)*
Time spent in a day to fetch water	4 hours	2.5 hours
Time spent in domestic water related activities*	1.5 hours	2 hours
Time spent in cattle washing activities	1.2 hours	1.2 hours
Total time spent in water related activities	6.5 hours	5.5 hours

*Cleaning utensils, washing clothes, etc.
Source: Primary Data (2012).

Table 8.2:
Responsibility for water-related tasks

	Potable Water Villages		*Saline Water Villages*	
	*Drinking Water**	*Livestock-Related Tasks^*	*Drinking Water**	*Livestock Related Tasks^*
Number of households in which males are involved	1	5	1	16
Number of households in which females are involved	161	155	140	121

*Fetching and storing water.
^Washing and drinking purposes.

location of water sources in both types of villages. The less time taken in saline water villages is due to the fact that in most cases, the purchased water is stored within the compound of the house. The eventual impact of distance of water source and the time spent in fetching enough water to fulfil the water requirement of a family is on the individual responsible for procurement of water.

As is evident from the literature, women and girls in the family are largely placed with the responsibility of fetching water. Responses obtained from the sampled population also resonate with the fact (Table 8.2). Of the total population involved in water related activities related to drinking and livestock, 96 per cent are females across both the village types. Among the women, 88 per cent women above 6 years are involved in water-fetching activities.

A bifurcation of village type, however, indicates that men in saline villages participate more actively in water activities related to livestock rearing. While in potable water villages, 2 per cent men engage in water activities for livestock, in case of saline water villages, this percentage rises to 6. Discussions with men and women reveal that engagement of men in water related activities is higher in saline water villages as the water for drinking purposes has to be extracted from the underground water tank. Since women are largely responsible for pulling out water for cleaning, cooking and drinking purposes, men contribute labour for extracting water for livestock. In case of potable water villages, the water sources are largely outside the compound of the house. Here, women are the ones responsible for procuring water and men do not engage as much in activities related to livestock.

8.2.2 Decision-making in Water Related Issues: Gender Roles

Literature suggests that the onus of decision-making regarding the construction, usage, and maintenance of water sources usually lies in the hands of the men folk (IANGWE, 2004). This is especially true in rural settings where confinement of women to only domestic household tasks is profound. Thus, in a region like Mewat where women are not allowed to move outside their homesteads without the permission of the elderly or the males in the house, the absence of women's participation in water-related issues is expected to be skewed. The analogy presented emerges to be true with 99 per cent households mentioning that it is the man in the household who decides the choice of water source for domestic, livestock and agricultural purposes. With respect to representation at the community level, it is usually the men in the village who engage in discussions. Therefore, men take decisions related to construction, maintenance and usage of community water sources. Women's participation in these domains is next to nil, which gets reflected in their knowledge of village water sources. Discussions with women and men separately reveal stark differences. While the men have an exact knowledge of the number of personal and underground water tanks in the village and the number of community water sources, this knowledge is completely missing among women. For instance, in Karheri village, while men informed us that there are 56 personal water tanks and two ponds in the village, women mentioned that only 10–15 personal water tanks and one pond existed. Lack of knowledge of community water sources among women has a direct implication on the time and effort women have to spend in procuring water or doing water-related activities. Women who are not aware of the second pond in the village depend on only one pond. Resultantly, during summer months, they face a lot of problem when water in that pond dries up.

Even though, men are the decision-makers in households, it is women who largely engage in procuring and managing household water requirements. The absence of women's stake in the decision-making process is bound to have an influence on their lives. With different roles that men and women play in a household, priorities are likely to differ. The differentiation reflects in the criteria on the basis of which

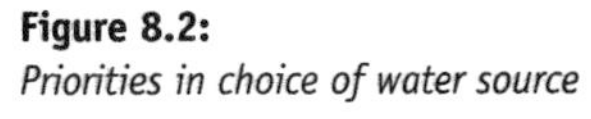

Figure 8.2:
Priorities in choice of water source

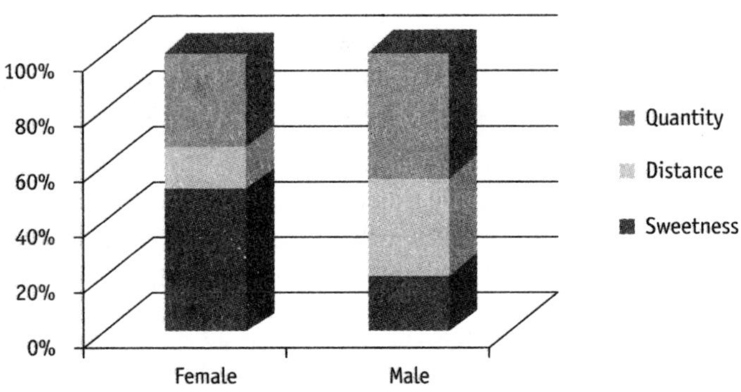

Source: Primary Data (2012).

men and women tend to choose water sources. While for men, distance of water source and the quantity are more of a priority in selecting a water source for domestic requirements, women concentrate more on the quality (Figure 8.2).

Discussions reveal that while women are keen to walk an extra kilometre to fetch water from a source which is cleaner, men are not very keen for women to do so. Owing to the cultural norms in the region, men are reluctant to send their wives and daughters out of the village to fetch water. Poverty-ridden households, which cannot afford personal water tanks (filled through tankers), are usually the ones who are forced to send their girls and women out of the village to fetch water from neighbouring villages. However, men tend to keep a tab on the movement of women and ensure that they move in groups instead of fetching water individually, especially when they go outside the village. Women's stress on quality water, however, is disregarded by men to an extent that women report instances of beating upon going to fetch water from a cleaner and a safer source. Box 8.1 details a case study of a village with saline water that has been drawn out from in-depth discussions with both, male and female inhabitants. It underlines the limited role played by women in taking decisions related to water management of the household despite them being solely responsible for the water-fetching exercise.

Box 8.1:
The plight of Karheri

Karheri is a small village located in the Nagina block of Mewat. Just about 2 km from the national highway, the village draws out a sad picture in contrast to its strategic location. The total number of households in the village range from 175 to 180 with pre-dominance of Meo Muslims. Poverty is evident across the village. Several households survive in temporary shacks and their inhabitants bathe only once a week due to dearth of water. They mention eating one or two meals a day as there is not enough food. Muddy roads with sewage water puddles are a common sight. Discussions with villagers revealed that the village headman ('*Sarpanch*') got the bricks extracted from the village paths and constructed the compound of his home. Other inhabitants witnessed the sight and succumbed to the situation without raising their voice. This experience is not new to them. The elderly narrate several incidents of past injustice. In this state, miseries of the inhabitants magnify when one of the most basic necessities of life, that is water, is scarce. The ground water in Karheri is saline to such an extent that an alkaline soap melts upon contact. The saline content is extreme due to which ground water cannot be used for livestock cleaning or for washing clothes.

Resultantly, agriculture is largely rain-fed. As far as domestic water requirements are concerned, the inhabitants are solely and individually responsible for meeting their needs. Even though, the village has piped water supply, it cannot be relied upon as the supply is erratic or there are no taps. The erratic nature of the supply is attributed to vested interests of the inhabitants of neighbouring villages from where the supply is sourced. They usually tend to block the supply pipes to ensure that water released from the government tanks is enough to meet their own needs. Consequently, villages like Karheri receive water only during months when demand for water is low such as during the rains or winters.

In this scenario, women who are largely responsible for domestic water management in a household face numerous challenges. With no stake in deciding the type and distance of the water source for procurement and working within cultural constraints, the drudgery of women is profound. The extent of suffering is also dependent on the socio-economic status of the household. If the house is affluent enough to afford a water tanker every month, the woman is largely entrusted with the task of pulling out water from the tank buried in the compound of the house. However, in impoverished households, a woman and her daughters, have to trudge a couple of kilometres one-way to fetch water from a chamber in the neighbouring village. A can of water estimated to weigh around 10 kg that she holds up on her head and walks. This exercise is repeated at least thrice in a day to fend for daily water requirement of the household and is routine. There are also some households that allow women to fetch water from a neighbouring village well. Comparatively, these women have to undergo less effort, as the well is closer. However, not every woman goes there. The decision to opt for the well or the chamber largely lies with the men in the household. Men, who are fine with their women negotiating with men of neighbouring village to procure water, opt for the well. The more conservative ones send their women to the water chamber that is a couple of kilometres away. The woman is only a puppet. The husband or the elderly in the household decide her actions. Differing priorities regarding choice of water source further magnify the plight of women who are also responsible for looking after the livestock. While women prefer a cleaner water source, men prefer distance. The disparity in opinion eventually results in women abiding by the instructions given by the men in the household. Non-adherence results in conflict, and at times physical violence is meted out to women.

Women have an important role to play in ensuring the health and well-being of household members. However, they have no stake in decision-making regarding any issue. Water is one such issue. The lack of women's participation in choice of water source magnifies the drudgery they have to undergo. Women in Karheri village suffer from several health issues. Majority of them attribute it to the physical agony they undergo in fetching water from long distances. While several report spine problems, some have undergone bouts of depression on account of carrying heavy loads on their heads. Girls mention being out of school only because they have to accompany their mothers in the water-fetching exercise. Education is not a priority for them. All these things coupled together highlight the misery of a *'Meoni'* (a Meo woman as denoted in the local dialect).

8.2.3 Socio-economic Impact of Differing Water Situations

The time and effort put into procuring and managing water impact the socio-economic status of the household. In case of potable water villages, women spend a lot of time in fetching water. Having a water source within the village, they are not allowed to go outside the village to fetch water from neighbouring sources. Resultantly, women in the village queue up for hours to procure water for the household; standing for hours together only to procure water leaves them with no time to rest. Often, water-fetching tasks are undertaken after other household duties are over. Standing for long hours results in back problems. Some also face spinal swelling due to carrying heavy loads on their heads. Furthermore, quality of water has to

Table 8.3:
Incidence of diseases

Village Type	*Incidences of Water-borne Diseases[5]* (per cent)*
Potable water villages	5.8
Saline water villages	8.8

*Statistically significant at 0.05 per cent level of confidence.

be compromised upon due to the decision taken by men regarding choice of water source. As a consequence, the quality of water procured for drinking purposes is not as good. Being illiterate and ignorant of household water purification practices, the same water is used for drinking without any treatment. Consequently, the incidence of water-borne diseases is too high. Incidence of water-borne diseases is higher in saline water villages as compared to potable water villages (Table 8.3).

The high incidence is indicative of increased expenditure on health-related issues involving treatment of diseases contracted by household members. The data reveals that amongst the non-purchasers, only 10 per cent households across both potable and saline water villages opt for a different source for drinking and other purposes. Rest 90 per cent use the same source for all purposes. This has a direct implication on the quality of water being consumed. Given that women largely are aware of the quality of water source, if they are the sole decision-makers, they can pick and choose the source accordingly, which can help reduce the incidence of water-borne diseases to a great extent.

Another implication of water scarcity is reflected in the lack of women and girls' education owing to the long hours they have to spend in water-related tasks. The data reveal that of the total number of girls between the ages of 6 and 14 years, which is considered as the school-going age, several accompany their mothers to fetch water. A village wise distinction however reveals stark facts with more number of girls from potable water villages accompanying others to fetch water than the number of girls in saline water villages. This has direct implication on the number of girls missing out on their education. One of the plausible reasons behind an increased number of girls engaged in water fetching in potable villages is that none of the households in potable water villages purchase water.

Resultantly, they have to move out to procure water from external sources. Contrastingly, in saline water villages, majority of households purchase water resulting in reduced number of girls in the school-going category engaged in water procurement activities.

Mostly, adolescent girls are pulled out of school after primary or secondary schooling to accompany their mothers to fetch water for the households. Girls belonging to poor households are engaged in either looking after livestock and washing clothes, or fetching water. Resultantly, they are unable to get adequate or any education at all. Lack of education has eventual implications as discussed earlier on health and hygiene practices followed by children and adults in a household. All these factors have long-term impacts on the socio-economic status of households.

8.3 Conclusion

This chapter captured the way of life of people residing in the water scarce region of Mewat, a remotely located district in Haryana state. It laid a special focus on the extent and type of women's participation in water issues and their related implications. Differing water situations were studied to understand the corresponding impact on women in particular and the households at large. Lack of water availability was found to have serious and direct implications on individual well-being. For regions like Mewat where water scarcity is an all-encompassing problem, procuring water for daily activities becomes a major challenge. And finally, the onus of procurement is on women and girls.

The two differing water situations in Mewat villages have different impacts on the lives of inhabitants. In potable water villages, the sources of water for drinking and other purposes are diverse and located within the village. In case of saline water villages, the inhabitants largely rely on purchasing water, which they store or walk long distances to other villages to fetch water. Government water sources in saline water villages are found to be dysfunctional due to the apathy from the government officials and resigned attitude of the villagers. However, against logical assumptions, the average time taken and effort undergone by a woman in procuring and managing water for a household is found to be more in the case of potable water villages. This is primarily due to the fact that women in potable water villages have to queue up for hours to fetch water from a village source. Contrastingly, women in saline water villages have to pull out water from the tank buried in the village compound. Overall, the economic implications for all households who can afford a personal source are very high. The less affluent cannot afford it and largely rely on donation from the rich or on physical effort.

One of the main findings is that besides water being a major element of survival, women are indispensible to water issues. While this may be true for other regions as well, the paradox of Mewat is unique. Even though, women are the ones who look after the entire water management of a household from procuring to using water sources, they have a grim presence in the decision-making scenario. It is the men who dominate the decisions related to construction, maintenance and usage of community water sources. The dominance of males is grave to such an extent that even within the village, the movement of women is restricted. Not being a part of discussions, women tend to be unaware of sources existing within the village. Resultantly, they depend on single sources. When these sources dry up, it leads to problems. Furthermore, the differential priorities among men and women regarding choice of water sources result in conflict and have socio-economic implications. While men stress on the distance of water source, women give priority to quality of water. Owing to the local culture, men prefer to send their women to sources, which do not require negotiations for usage. Resultantly, abiding by the instructions given by men, women tend to procure water from sources that may not be as clean. This results in higher incidence of water-borne diseases. The same has economic implication with respect to spending more on the treatment. Having large families, the water requirements are high. As a consequence, households who do not have access to water within the village traverse kilometres together to procure water. To reduce the burden on one woman in the household, girls accompany their mothers and sisters for the water-fetching exercise. As a result, they do not attend school and hence, receive no education.

It is imperative to upscale the participation of women in decisions related to water management at the community level. The drudgery that they undergo is massive, which has implications not only on their health but the entire family. Through the facts presented in this chapter, it becomes clear that water is one of the major reasons for pulling people into poverty in Mewat region. A large portion of their income goes into either purchasing water or digging borewells and tube wells. Furthermore, females in the region cannot receive education due to their occupation of fetching and managing water for the household. Being out of education, they remain oblivious of using techniques to purify water, better management in case of scarcity and effective use of resources. All these factors further deteriorate the situation of women as well as the household. As women are the forbearers of water management, the situation calls for a more active role of women in the decision-making of construction, maintenance and usage of water sources. Going beyond this avenue, there arises a need to change the patriarchal values among the inhabitants so that empowerment of women can be facilitated by the men as well.

Notes

1. *Aravalis* is a range of fold mountains in Western India spread across 800 km in north eastern direction. It cuts across Indian states of Gujarat, Rajasthan, Haryana and Delhi. It is also called Mewat hills locally (Wikipedia, 2014a).
2. Meo Muslims are Muslim Rajputs tracing their roots to North Western India spread across Rajasthan, Haryana and Uttar Pradesh. They profess the beliefs of Islam but their ethnic structure is similar to the Hindu caste society (Wikipedia, 2014b).
3. INR to US$ conversion on 24 April 2013 from http://themoneyconverter.com/INR/USD.aspx
4. The figures have been arrived at by capturing information on parameters such as distance of water source, number of people engaged in water-related exercise and number of times a person has to attend to the water-related task.
5. Water-borne diseases captured in the study include diarrhea, jaundice, typhoid, gastroenteritis and skin allergies.

References

Aureli, A. and Brelet, C. 2004. *Women and Water: An Ethical Issue*. Retrieved from http://unesdoc.unesco.org/images/0013/001363/136357e.pdf (accessed on 25 September 2012).

Census of India. 2011. *Provisional Population Data*. 2011. Retrieved from www.census2011.co.in (accessed on 8 July 2011).

DESA: United Nations Department of Economic and Social Affairs. 2010. *The Millennium Development Goals Report*. Retrieved from http://www.un.org/millenniumgoals/pdf/MDG%20Report%202010%20En%20r15%20-low%20res%2020100615%20-.pdf#page=60 (accessed on 1 October 2012).

Food and Agriculture Organization (FAO). 2000. Retrieved from http://fao.org/gender/gender.htm (accessed on 1 October 2012).

Interagency Network on Gender and Women Equality (IANGWE). 2004. *A Gender Perspective on Water Resources and Sanitation*. Background Paper No. 2. United Nations Department of Economic and Social Affairs. Retrieved from http://www.unwater.org/downloads/bground_2.pdf (accessed on 23 April 2013).

International Atomic Energy Agency (IAEA). 2007. *Women and Water*. Retrieved from http://www.iaea.org/newscenter/news/2007/womenday2007.html (accessed on 25 September 2012).

Mehta, P.K., Saxena, N. and Kumar, A. 2011. *Impact Assessment of IRRAD's Interventions in Select Villages of Mewat*. Rural Research Center, Institute of Rural Research and Development, Gurgaon, Haryana.

Nash, K. 2010. *Women and Water Rights: Rivers of Regeneration*. Retrieved from http://womenandwater.net/ (accessed on 1 October 2012).

National Commission for Women (NCW). 2005. *Women, the Water Providers*. Retrieved from http://ncw.nic.in/pdfreports/Women%20&%20Water.pdf (accessed on 23 April 2013).

UNEP. 2012. *Women and Water Management: An Integrated Approach*. Retrieved from http://www.unep.org/pdf/women/ChapterFive.pdf (accessed on 1 October 2012).

Wikipedia. 2014a. *Aravalli Range*. Retrieved from http://en.wikipedia.org/wiki/Aravalli_Range on 16 August 2014.

Wikipedia. 2014b. *Meo (Ethnic Group)*. Retrieved from http://en.wikipedia.org/wiki/Meo_(ethnic_group) on 16 August 2014.

9

The High Fluoride Burden and Tribal Women: Occurrence and Remedy

Tapas Chakma, Gregor von Medeazza, Sanjay Singh and Pradeep Meshram

9.1 Introduction

Fluorosis[1] has emerged as an important public health problem in India. It affects multiple body organs and systems. Its clinical manifestations start from damaged and discoloured teeth and ends in crippling conditions (Teotia and Teotia, 1994). Dental fluorosis is characterised by permanent hypo-mineralisation (Mullins et al., 1998; Murray and Wilson, 1948). It is estimated that in India, over 18 million people are affected by dental fluorosis and nearly 8 million people are affected by skeletal fluorosis (Fewtrell et al., 2006). Most of the eastern, southern and western districts in the Indian state of Madhya Pradesh have excess fluoride in ground water.[2] Madhya Pradesh is one of the largest states, located in central India with a population of over 72 million (Government of India, 2012).

The majority of the fluoride-affected areas are heavily inhabited with Madhya Pradesh's tribal population. These groups have traditionally been economically and socially more vulnerable than other social groups in India. Madhya Pradesh's tribal groups constitute over 20 per cent of the state's population and are mainly concentrated in the southern belt of the state. Due to hydro-geomorphological reasons, most of the districts located in this region feature high fluoride concentrations in their groundwater, which in some places, exceed 13 ppm (PHED and UNICEF, 2007), 1.5 ppm being the national permissible limit. Besides, dental and skeletal fluorosis, which are the medical conditions the present chapter focuses on, evidence gathered indicated that fluoride contamination has also have other grave but invisible implications. Indeed, results from a study conducted by Saxena et al. (2012) in Madhya Pradesh showed that children in endemic areas of fluorosis are at risk of impaired cognitive development. This means that fluoride contamination has even more profound implications on society than presently recognised.

The number of the contaminated sources varies from district to district. The adjacent districts of Seoni and Mandla, situated in the south-eastern part of the state, are known for their endemic fluorosis. Geographically and socio-economically both the districts are similar and their main sources of drinking water are drawn from the aquifer through hand pumps. Over the last 15 years, UNICEF has been supporting several investigations carried out by the Regional Medical Research Centre for Tribals (RMRCT), Jabalpur to shed light on the prevalence of severe forms of dental and skeletal fluorosis (for Mandla district, see for instance Chakma et al., 1996, 2000). Clinical and anthropogenic examination in fluoride-affected villages indicated that the women are more severely affected by skeletal fluorosis than men, even when subject to similar quantities of fluoride contamination in their drinking water. This is because of the poor socio-economic conditions of these tribal women and girls, linked their less access to the required micronutrients needed to mitigate the excess fluoride intake from drinking water (Rao et al., 2010). Furthermore, previous RMRCT's UNICEF-supported research in 2006 established that, in adults, more than 40 per cent of the fluoride intake comes from food intake (Godfrey et al., 2011); dietary modifications are thus essential for fluorosis mitigation, and form an essential component of a sound Integrated Fluoride Mitigation programme (NEERI, 2007). This 40 per cent food intake proportion has to be understood both as a source of fluoride from certain foodstuff rich in fluoride (such as tea and tobacco), but also as a way to mitigate the excess fluoride in the human body through other foodstuff, rich in vitamin C and calcium such as the green vegetable, Cassia tora (Godfrey et al., 2011). Chakma et al. (2000) reported presence of fluoride in 70 different commonly consumed food items and habits (such as tobacco chewing) in Mandla district of Madhya Pradesh. In their study, nutrition supplementation has emerged as the essential key factor for fluorosis mitigation.

Considering these facts, a subsequent intervention study was undertaken in Seoni district by RMRCT with support from UNICEF between 2010 and 2012. The main focus of the study was to ensure adequate amounts of calcium, vitamin C and iron were provided to fluoride-affected communities, especially women. This was done through dietary modification and the introduction of easily available local foodstuff, rich in these micronutrients. Women were kept at the centre of the intervention. They were the key actors in terms of knowledge dissemination, and change agents in the dietary habits of the family members in the view of reducing the incidences of fluorosis. This chapter presents the findings of the survey. It shows how women are more significantly affected by fluorosis, and argues that women can potentially lead the way to improve the situation through specific interventions.

9.2 Methodology

The cross-sectional intervention study was carried between 2010 and 2012 in Dhanora and Seoni blocks of Seoni district of Madhya Pradesh (RMRCT, 2012). Seoni district was purposely selected, as this is one of Madhya Pradesh's fluoride worst affected districts, with no perennial rivers to supply alternative drinking water. Medical officers, biochemists, nutritionists, medical social workers and trained field workers were involved during the entire study using standard equipment and procedures.

9.2.1 Selection of Village and Households

Based on the fluoride concentration data collected by the Public Health Engineering Department (PHED) in Seoni block, 64 villages were found to have the greatest risk of fluorosis. In these villages, more than 50 per cent of drinking water sources were found contaminated with high fluoride levels (i.e. with fluoride levels above 2 mg/l). From these 64 villages, a total of 10 villages were randomly selected. The affected villages were stratified according to the proportion of fluoride-contaminated hand pumps. The study was conducted in two blocks of Seoni district (in Madhya Pradesh), chosen on the basis of highest proportion of hand pumps contaminated with high concentrations of fluoride (ranging from 2 to 13 mg/l).

Within the selected village, all households and available individuals were covered for the study. Total 5,437 individuals, 2,641 males and 2,796 females, of different age groups were surveyed at the time of the baseline in 2010. Out of them, 5,037 individuals (2,496 males and 2,541 females) were still available at the time of the post-intervention evaluation in

Table 9.1:
Age and sex distribution of study population

Age Group (years)	Pre-intervention			Post-intervention		
	Male	Female	Total	Male	Female	Total
≤6	207	216	423	170	149	319
7–14	612	637	1,249	494	497	991
15–19	265	293	558	316	311	627
20–29	442	511	953	421	459	880
30–39	392	417	809	362	417	779
>40	723	722	1,445	733	708	1,441
Total	2,641	2,796	5,437	2,496	2,541	5,037

2012 (Table 9.1). Seventy-two water samples from drinking water sources were analysed during the baseline. We analysed 345 urine samples pre-intervention and 200 urine samples post-intervention.

9.2.2 Data Collection

During the pre- and the post-intervention periods, a complete clinical examination was done. All the 5,437 adults were asked to perform standard exercises shown to ascertain the muscular stiffness due to fluorosis.

All 72 drinking water sources were tested and their respective fluoride concentrations were determined. Spot urine samples of individuals showing signs of skeletal abnormalities as well as early warning signs were analysed for fluoride by standard testing methods using an ion selective electrode (ISE) metre. Finally, a dietary survey was conducted using the 24-hour recall method (Thimmayamma, 1981).

9.2.3 Data Analysis

Data were entered into a specially developed programme using *CS pro* 4.0 version software. Double data entries were done as a procedure to check for data entry errors. Using *SPSS Windows* version 17.0, Univariate analysis was undertaken to determine the prevalence of various form of fluorosis in different age groups. A cross-tabulation was performed to determine any association of pre- and post-intervention. Statistical *t*-test was applied to determine the significance levels.

9.3 Intervention Measures: Integrated Fluoride Mitigation

The baseline survey found the prevalence of skeletal fluorosis to vary between 0 and 6 per cent amongst the different age groups. The prevalence of non-skeletal fluorosis, especially the gastrointestinal symptoms, varied between 5 and 14 per cent. It was also observed that the cohort's diet was grossly deficient of several micronutrients, especially calcium, vitamin C and iron. An Integrated Fluoride Mitigation intervention strategy was developed and implemented accordingly. This strategy was based on a two-pronged approach to address fluorosis: firstly, reducing fluoride intake by providing safe fluoride-free drinking water and reducing the consumption of certain foodstuff; secondly, increasing the intake of certain micronutrients and other beneficial foodstuff to mitigate the excess and effects of fluoride in the body.

Every intervention village was mapped to locate safe drinking water sources and each source was labelled as 'safe' and 'unsafe'. Villagers were guided to use only the safe source to draw their water for drinking and cooking. A team of female investigators was placed in each of villages and a detailed Information Education and Communication (IEC) plan was developed, especially targeting the principal female member of each family. The underlying objective of the IEC approach was also to establish a rapport with the female members of the community and to identify natural leaders who could later play a crucial role as agents of change in the project. In this view, the trained field investigators visited the selected villages to organise group meetings. They created awareness on the importance of drinking safe water and the inclusion of nutritional supplements into dietary routines. The objective of the dietary modification component was to decrease the amount of fluoride intake (by reducing consumption of tea and tobacco) while increasing the intake of foodstuff rich in calcium, vitamin C and iron to mitigate excess fluoride in the body.

In one of the 10 villages, as there was not a single safe drinking water source, a community-based photovoltaic electrolytic defluoridation (EDF) plant was installed by the National Environmental Engineering Research Institute (NEERI). This was done in collaboration with UNICEF and Madhya Pradesh's Government PHED. This defluoridation unit enabled the community to access safe drinking water, during and beyond the survey/intervention presented in this chapter. This EDF plant was the first unit to be installed in Madhya Pradesh, and became a model of viable community-managed defluoridation, now replicated by PHED with support from NEERI and UNICEF in other districts across the state.

9.4 Results

9.4.1 Dental, Skeletal and Non-skeletal Fluorosis

The overall prevalence of dental fluorosis[3] was observed among 1,032 individuals (18.9 per cent of the studied sample) during the baseline survey and reduced to 823 individuals (17.6 per cent) in the end line survey after the intervention took place. Total 51.9 per cent of dental fluorosis sufferers were children and young individuals under the age of 20 years; females were more affected than males. After the intervention, dental fluorosis among children and youth reduced significantly from 49 per cent to 36 per cent (Table 9.2).

Skeletal fluorosis[4] was also measured. Pre- and post-intervention prevalence of genu valgum (knocked knee) was also slightly higher in women (9.4 per cent) than in men (9.3 per cent), and was significantly reduced to 6.8 per cent for both men and women after the intervention (Table 9.3). At the end of the intervention, no new case was reported in all the 10 studied villages.

It is important to emphasise that there were seven cases of severe crippling fluorosis recorded as part of this study and all of them were females and the most in the reproductive age group. Skeletal fluorosis did not affect any man.

Table 9.2:
Percentage distribution of dental fluorosis during pre- and post-intervention periods

	Pre-intervention				*Post-intervention*			
	Male		*Female*		*Male*		*Female*	
Age Group (years)	*No. Examined*	*No. with Dental Fluorosis (per cent)*	*No. Examined*	*No. with Dental Fluorosis (per cent)*	*No. Examined*	*No. with Dental Fluorosis (per cent)*	*No. Examined*	*No. with Dental Fluorosis (per cent)*
<6	207	0	216	0	170	0	149	0
7–14	612	284 (46.4)	637	316 (49.6)	494	210 (42.5)	497	217 (43.7)
15–19	265	137 (51.7)	293	153 (57.7)	316	135 (42.7)	311	141 (45.3)
20–29	442	87 (19.7)	511	55 (10.8)	421	100 (23.8)	459	60 (13.1)
30–39	392	0	417	0	362	9 (2.5)	417	19 (4.5)
>40	723	0	722	0	733	4 (0.5)	708	10 (1.3)
Total	2,641	508 (19.2)	2,796	524 (18.7)	2,496	471 (18.9)	2,541	417 (16.4)

Table 9.3:
Percentage of genuvalgum (GV) prevalence during pre- and post-intervention periods

Age Group (years)	Pre-intervention				Post-intervention			
	Male		Female		Male		Female	
	Number Examined	Number with GV and %	Number Examined	Number with GV and %	Number Examined	Number with GV and %	Number Examined	Number with GV and %
<6	207	16 (7.7%)	216	15 (6.9%)	170	6 (3.5%)	149	12 (8.1%)
7–14	612	164 (26.8%)	637	192 (30.1%)	494	96 (19.4%)	497	104 (20.9%)
15–19	265	36 (13.6%)	293	40 (13.7%)	316	41 (12.9%)	311	38 (12.2%)
20–29	442	23 (5.2%)	511	9 (1.8%)	421	17 (4.0%)	459	11 (2.4%)
30–39	392	3 (0.8%)	417	3 (0.7%)	362	6 (1.7%)	417	3 (0.7%)
>40	723	4 (0.6%)	722	4 (0.6%)	733	3 (0.4%)	708	1 (0.1%)
Total	2,641	246 (9.30%)	2,796	263 (9.4%)	2496	169 (6.8%)	2,541	174 (6.8%)

GV, genu valgum.

Figure 9.1:
Exercises to ascertain muscle stiffness due to fluorosis

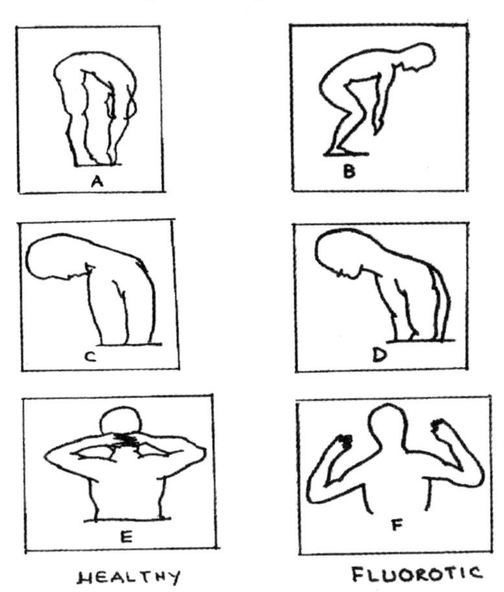

Source: Based on Susheela (2001).

Furthermore, the prevalence of skeletal fluorosis observed through the standard procedures (refer to Figure 9.1) such as the inability of the surveyed individual in touching their chest with the chin, bending forward, performing sit-ups and touching the back of their head with both palms and shoulders extended, was recorded in 3.6 per cent, 1.5 per cent, 1.1 per cent and 1.6 per cent of women, respectively. These skeletal fluorosis indicators were higher for women than those recorded for the male category, which were 3.1 per cent, 0.5 per cent and 1.1 per cent, respectively. Similar results of women being more affected than men were also previously reported by Susheela et al. (2010). After the intervention, these same indicators respectively reduced to 2.8 per cent, 0.7 per cent and 1.1 per cent for the women group in the span of 1 year (Tables 9.4 and 9.5). The reduction is statistically significant ($P < 0.05$) in the case of the indicator 'touching chest with chin' and 'bending forward'.

Finally, non-skeletal fluorosis[5] was measured. Nearly 54 per cent of the surveyed individuals were suffering from gastrointestinal disturbance,

Table 9.4:

Distribution of signs of skeletal fluorosis during pre- and post-intervention periods

	Pre-intervention				Post-intervention				Pre-intervention				Post-intervention			
	Touching Chest with Chin				Touching Chest with Chin				Bending Forward				Bending Forward			
	Male		Female		Male		Female		Male		Female		Male		Female	
Age Group (years)	No. Examined	No. Unable to Do (per cent)	No. Examined	No. Unable to Do (per cent)	No. Examined	No. Unable to Do (per cent)	No. Examined	No. Unable to Do (per cent)	No. Examined	No. Unable to Do (per cent)	No. Examined	No. Unable to Do (per cent)	No. Examined	No. Unable to Do (per cent)	No. Examined	No. Unable to Do (per cent)
<6	207	0	216	0	170	0	149	0	207	0	216	0	170	0	149	0
7–14	612	12 (2.0)	637	17 (2.7)	494	2 (0.4)	497	5 (1.0)	612	22 (3.6)	637	16 (2.5)	494	5 (1.0)	497	4 (0.8)
15–19	265	10 (3.8)	293	11 (3.8)	316	3 (0.9)	311	6 (1.9)	265	7 (2.6)	293	6 (2.0)	316	2 (0.6)	311	3 (1.0)
20–29	442	13 (2.9)	511	10 (2.0)	421	10 (2.4)	459	9 (2.0)	442	2 (0.5)	511	4 (0.8)	421	1 (0.2)	459	1 (0.2)
30–39	392	15 (3.8)	417	22 (5.3)	362	12 (3.3)	417	19 (4.6)	392	4 (1.0)	417	2 (0.5)	362	3 (0.8)	417	1 (0.2)
>40	723	32 (4.4)	722	42 (5.8)	733	24 (3.3)	708	32 (4.5)	723	15 (2.1)	722	14 (1.9)	733	10 (1.4)	708	11 (1.6)
Total	2,641	82* (3.1)	2,796	102** (3.6)	2,496	52* (2.1)	2,541	70** (2.8)	2,641	50# (1.9)	2,796	42## (1.5)	2,496	21# (0.8)	2,541	20## (0.8)

* $t = 1.537$, df = 5,335, $P > 0.05$; # $t = 3.395$, df = 5,135, $P < 0.01$; ## $t = 2.378$, df = 5,335, $P < 0.05$.

Table 9.5:
Distribution of signs of skeletal fluorosis during pre- and post-intervention periods

Age Group (years)	Pre-intervention Sit-ups Male No. Examined	Pre-intervention Sit-ups Male No. Unable to Do (per cent)	Pre-intervention Sit-ups Female No. Examined	Pre-intervention Sit-ups Female No. Unable to Do (per cent)	Post-intervention Sit-ups Male No. Examined	Post-intervention Sit-ups Male No. Unable to Do (per cent)	Post-intervention Sit-ups Female No. Examined	Post-intervention Sit-ups Female No. Unable to Do (per cent)	Pre-intervention Touching Back of Head Male No. Examined	Pre-intervention Touching Back of Head Male No. Unable to Do (per cent)	Pre-intervention Touching Back of Head Female No. Examined	Pre-intervention Touching Back of Head Female No. Unable to Do (per cent)	Post-intervention Touching Back of Head Male No. Examined	Post-intervention Touching Back of Head Male No. Unable to Do (per cent)	Post-intervention Touching Back of Head Female No. Examined	Post-intervention Touching Back of Head Female No. Unable to Do (per cent)
<6	207	0	216	0	170	0	149	0	207	0	216	0	170	0	149	0
7–14	612	2 (0.3)	637	6 (0.9)	494	3 (0.6)	497	2 (0.4)	612	10 (1.6)	637	8 (1.3)	494	2 (0.4)	497	3 (0.6)
15–19	265	0	293	1 (0.3)	316	0	311	3 (1.0)	265	5 (1.9)	293	6 (2.0)	316	3 (0.9)	311	2 (0.6)
20–29	442	1 (0.2)	511	1 (0.2)	421	0	459	2 (0.4)	442	2 (0.5)	511	3 (0.6)	421	3 (0.7)	459	2 (0.4)
30–39	392	0	417	4 (1.0)	362	1 (0.3)	417	1 (0.2)	392	1 (0.3)	417	8 (1.9)	362	0	417	6 (1.4)
>40	723	11 (1.5)	722	19 (2.6)	733	9 (1.2)	708	10 (1.4)	723	11 (1.5)	722	20 (2.8)	733	8 (1.1)	708	15 (2.1)
Total	2,641	14 (0.5)	2,796	31* (1.1)	2,496	13 (0.5)	2,541	18* (0.7)	2,641	29# (1.1)	2,796	45## (1.6)	2,496	16# (0.6)	2,541	28## (1.1)

* $t = 1.537$, df = 5,335, $P > 0.05$; # $t = 1.943$, df = 5,135, $P > 0.05$; ## $t = 1.574$, df = 5,335, $P > 0.05$

out of which the majority (61.3 per cent) were women. Similarly, the prevalence of gastrointestinal manifestations in women, such as loss of appetite (13.9 per cent), gas in stomach (14.9 per cent), pain in stomach (9.7 per cent), chronic diarrhoea (7 per cent) and chronic constipation (6.5 per cent) was considerably higher at baseline among the studied women population than in men. As a result of the intervention, gastrointestinal manifestations in women significantly reduced to 6, 9.6, 2.3, 1.1 and 5.9 per cent, respectively (Table 9.6). It must be noted that the above listed symptoms are fairly common. Since they are also associated with several other diseases, they make it difficult for physicians to diagnose fluorosis.

Finally, the prevalence of other symptoms of non-skeletal fluorosis was also higher in women than in men: chronic headache (19.5 per cent), frequent tendency to urinate (12.6 per cent) and extensive weakness (18.6 per cent) were observed. Again, as a result of the intervention, these symptoms significantly reduced in women to 8.1 per cent, 2.5 per cent and 7.8 per cent, respectively (Table 9.7).

9.4.2 Biochemical Parameters

Among the biochemical parameters we analysed, only water fluoride and urine fluoride level are relevant to fluorosis. A total of 72 water sources were identified in the 10 study villages. Out of these, 31 (43 per cent of total) sources were contaminated with high fluoride (>1.5 ppm), while 39 sources were found to be safe, as per national standards. The fluoride value ranged from 1.5 ppm to 11 ppm. The symptomatic baseline prevalence of urinary fluoride (>2 ppm) was 91.9 per cent; while in post-intervention, it was reduced to 67%. This reduction was found to be statistically significant ($P < 0.001$).

9.5 Conclusion

For the past two decades, fluorosis has been endemic in Madhya Pradesh. Mainly due to the over-exploitation of groundwater, there has been a sharp increase in fluoride content. As a consequence, the number of fluorosis

Table 9.6:

Percentage prevalence of gastro-intestinal manifestations during pre- and post-intervention periods

Variables	Total Population Surveyed	Loss of Appetite		Gas in Stomach		Pain in Stomach		Diarrhoea		Constipation	
		Male (numbers and per cent)	Female (numbers and per cent)	Male (numbers and per cent)	Female (numbers and per cent)	Male (numbers and per cent)	Female (numbers and per cent)	Male (numbers and per cent)	Female (numbers and per cent)	Male (numbers and per cent)	Female (numbers and per cent)
Pre-intervention	Male N = 2,641 Female N = 2,796	243 (9.2)	390 (13.9)	312 (11.8)	417 (14.9)	379 (6.9)	530 (9.7)	144 (5.5)	195 (7.0)	130 (4.9)	183 (6.5)
Post-intervention	Male N = 2,496 Female N = 2,451	113 (4.5)	167 (6.6)	209 (8.4)	243 (9.6)	198 (3.9)	115 (2.3)	23 (0.9)	29 (1.1)	128 (5.1)	150 (5.9)
t value		6.636	8.610	4.033	5.807	4.737	11.054	9.272	10.577	0.329	0.898
df		5,135	5,245	5,135	5,245	5,135	5,245	5,135	5,245	5,135	5,245
P		<0.0001	<0.0001	<0.001	<0.001	<0.001	<0.0001	<0.0001	<0.0001	>0.05	>0.05

Table 9.7:

Percentage prevalence of other symptoms of fluorosis during pre- and post-intervention periods

		Other Symptoms					
		Headache		*Frequent Tendency to Urinate*		*Extensive Weakness*	
Variables	*Total Population Surveyed*	*Male (numbers and per cent)*	*Female (numbers and per cent)*	*Male (numbers and per cent)*	*Female (numbers and per cent)*	*Male (numbers and per cent)*	*Female (numbers and per cent)*
Pre-intervention	Male N = 2,641 Female N =2,796	283 (10.7)	544 (19.5)	244 (9.2)	352 (12.6)	245 (9.3)	521 (18.6)
Post-intervention	Male N = 2,496 Female N = 2,451	110 (4.4)	207 (8.1)	53 (2.1)	63 (2.5)	94 (3.8)	198 (7.8)
t value		8.496	11.812	10.925	13.546	7.920	11.402
df		5,135	5,245	5,135	5,245	5,135	5,245
P		<0.0001	<0.0001	<0.0001	<0.0001	<0.0001	<0.0001

cases has increased in recent times. In this study, crippling fluorosis (the most visible, dramatic and painful form of fluorosis) was only observed in women. It was also observed that the prevalence of skeletal fluorosis, gastrointestinal disturbance and other symptoms of non-skeletal fluorosis were significantly higher in women than in men. These indicate that fluoride causes extra burden to physiology whenever calcium demand increases. Furthermore, symptoms of skeletal fluorosis cases were seen more prevalently in adults, as these symptoms worsen with age. Skeletal fluorosis symptoms (such as the inability of touching one's chest with the chin and touching the back of one's head with both hands) are more prevalent among women of the 30–39 years age group as compared to men. We also observed a higher prevalence of non-skeletal fluorosis (gastrointestinal manifestations) among females as compared to males. This is mainly because of the increase in the calcium demand of the body during women's reproductive age. Furthermore, due to poor nutritional intakes, especially in the poorer tribal areas, women almost always suffer from a calcium deficiency, which aggravates the fluorosis problem. It is generally believed that skeletal fluorosis (early signs) is an irreversible clinical condition.

However, the present study indicates that a reduction and even reversal of skeletal fluorosis (early signs) among women can be achieved through an Integrated Fluoride Mitigation intervention (RMRCT, 2012). Such intervention encompasses both a reduction in fluoride intake through safe drinking water and consumption of lower fluoride content foodstuff, as well as the mitigation of fluoride intake through increased consumption of calcium, vitamin C and iron-rich foodstuff. The dietary modification was a key element of the intervention, considering that around 40 per cent of the total fluoride intake is through foodstuff; adequate dietary modifications are thus essential for fluorosis mitigation.

In our intervention, women were the key change agents in terms of health education and improvements in household dietary habits. Well-trained field investigators organised daily group meetings with a special focus on womenfolk. Educational sessions were carried out, outlining the main issues and how fluorosis it can be mitigated through safe drinking water as well appropriate dietary intakes. The UNICEF-supported RMRCT study showed that women are the key stakeholders to ensure successful Integrated Fluoride Mitigation interventions (RMRCT, 2012). Especially in rural tribal communities, women are the main household members who decide from which source to fetch water and which ingredients to use for cooking. Women are also the decision-makers to modify the dietary habits of their entire family (RMRCT, 2012).

Given the societal barrier for men to access and change the habits of others' households, the involvement of women in this project has demonstrated that their leadership is central for fluorosis mitigation. Throughout the two year-long project, it has been observed how women emerged as natural leaders,[6] volunteering and undertaking concrete action in changing the dietary habits of their families and other households. In addition, they ensured that safe drinking water was obtained. The results highlight that Integrated Fluoride Mitigation interventions have a significant positive impact on communities living in fluoride affected areas. These positive impacts are especially visible and beneficial for women who on average suffer a greater burden from fluorosis. The UNICEF-supported RMRCT tested interventions resulted in a significant improvement of the main fluorosis parameters, entailing an overall enhancement of the quality of life of these women.

In conclusion, it is evident from the survey data analysis that women are more heavily impacted by fluorosis than men. This chapter has shown that reasons are multiple. Lower average nutritional levels and reproductive function, combined with their higher burden from household chores requiring strenuous effort while fetching water, all lead to greater vulnerability to fluorosis. Solutions have been successfully demonstrated over the last couple of years as part of the UNICEF-supported RMRCT programme. It has been shown that women play a leading role in implementing those solutions. As part of the Integrated Fluoride Mitigation strategy, the nutrition intervention programme focused primarily on women who are the key decision-makers when it comes to household modifications, in terms of water and food intake. Women's participation must be given primary importance in all community-led water intervention plans. The UNICEF-supported intervention undertaken by RMRCT in Seoni over the last couple of years clearly showed that women play a central role in significantly reducing the prevalence of fluorosis-related symptoms within their communities. They make Integrated Fluoride Mitigation approach effective, linking safe drinking water with nutritional supplementation.

The sectorial review carried out jointly by UNICEF, FAO and SaciWATERs on the situation and prospects of WASH in India (UNICEF, FAO and SaciWATERs, 2013) discussed the roles of major stakeholders in overcoming the current problems of the water sector. The review emphasised that each stakeholder has water responsibilities to honour, and that water issues cannot be solved without commitment and integrity. Similarly, the fluorosis issue cannot be solved unless all concerned stakeholders actively play their part, be it the community, the state, the market, or civil society. And in the case of Integrated Fluoride Mitigation, the traditional boundaries of the water sector need to be crossed, as its problem is complex and multi-sectorial, calling for greater convergence amongst key players not only in WASH but also nutrition, health and education.

Disclaimer

The views expressed herein are those of the authors and do not necessarily reflect the views of UNICEF or the United Nations.

Notes

1. The prolonged intake of water containing excess fluoride may result in the medical condition known as fluorosis.
2. www.mdws.gov.in
3. Dental fluorosis is a developmental disturbance of dental enamel caused by successive exposures to high concentrations of fluoride during tooth development, leading to enamel with lower mineral content and increased porosity (Alvarez et al., 2009). It is visible by the discoloration of the teeth from white to yellow, brown to black from the gum margin to the edge of the tooth.
4. Skeletal fluorosis weakens the bone structure and stiffens the muscles of affected individuals. The excess fluoride intake leads to bones losing some of their calcium, which then fixate on stiffening muscles. Skeletal fluorosis can have dramatic crippling effects with affected individuals suffering from muscle stiffness, knock knees, spinal bowing, arched legs, etc.
5. Non-skeletal fluorosis affects the body's organs and soft tissues and leads to ailments such as gastrointestinal mucosa and renal tubules.

6. 'Natural leaders' can be defined as community members who are enthusiastic and willing to take responsibility about making a positive change in their community. They are typically strong characters, able to influence their fellow community members and are supported by the majority of the community who are receptive to their views.

References

Alvarez, Jenny Abanto, Karla Mayra P.C. Rezende, Susana María Salazar Marocho, Fabiana B.T. Alves, Paula Celiberti and Ana Lidia Ciamponi. 2009. Dental fluorosis: Exposure, prevention and management. *Journal of Clin Exp Dent*, 1(1): 14–18.

Chakma, Tapas, S.B. Singh, S. Godbole and R.S. Tiwary. 1997. Endemic fluorosis with genuvalgum syndrome in a village of district Mandla, Madhya Pradesh, *Indian Pediatrics*, 34: 232–236.

Chakma, Tapas, P. Vinay Rao, S.B. Singh and R.S. Tiwary. 2000. Endemic fluorosis and other bony deformities in two villages of Mandla district, central India. *Fluoride,* 33(4): 187–195.

Fewtrell, L., S. Smith, D. Kay and J. Bartram. 2006. An attempt to estimate the global burden of disease due to fluoride in drinking water. *Journal of Water and Health*, 4(4): 533–542.

Godfrey, Sam, Pawan Labhasetwar, Tapas Chakma, Satish Wate, Aditya Swami and Jamie Bartram. 2011. Assessing and managing fluorosis risk in children and adults in rural Madhya Pradesh, India. *Journal of Water Sanitation and Hygiene for Development*, 1(2): 136–143.

Government of India. 2012. *Census of India 2011. Houses, Household Amenities and Assets. Latrine Facility*. Government of India.

Mullins, M., C. Warden and D. Barnum. 1998. Pediatric death and fluoride containing wheel cleaner (Letter). *Annals of Emergency Medicine*, 31: 523–525.

Murray, M.M. and D.C. Wilson. 1948. Fluorosis and nutrition in Morocco: Dental studies in relation to environment. *British Dental Journal*, 84: 97–100.

NEERI. 2007. *Integrated Fluorosis Mitigation: A Guidance Manual*. Nagpur: National Environmental Engineering Research Institute.

PHED and UNICEF. 2007. Rural Drinking Water Quality Atlas of Madhya Pradesh, prepared jointly by UNICEF, Public Health Engineering Department (PHED, Government of Madhya Pradesh), Regional Research Laboratory and Madhya Pradesh Council of Science and Technology, Bhopal.

Rao, Mallikarjuna K., N. Balakrishna, N. Arlappa, A. Laxmaiah and G.N.V. Brahmam. 2010. Diet and nutritional status of women in India. *Journal of Human Ecology*, 29(3): 165–170.

RMRCT. 2012. *Fluorosis mitigation, assessment, intervention and impact of intervention in fluorosis affected villages of Seoni district of Madhya Pradesh*, Report to UNICEF (unpublished).

Saxena, S., A. Sahay and P. Goel. 2012. Effect of fluoride exposure on the intelligence of school children in Madhya Pradesh. *Journal of Neurosciences in Rural Practice*, 3(2): 144–149.

Susheela, A.K. 2001. *A Treatise on Fluorosis*, 2nd ed. Fluorosis Research and Rural Development Foundation. New Delhi: India Publication.

Susheela, A.K., N.K. Mondal, Rashmi Gupta, Kamla Ganesh, Shashikant Brahmankar, Shammi Bhasin and G. Gupta. 2010. Effective interventional approach to control anemia in pregnant women. *Current Science*, 98(10): 1320–1330.

Teotia, S.P.S. and M. Teotia. 1994. Endemic fluorosis in India. A challenging National Health Problem. *Journal of Association of Physicians of India*, 32: 347–352.

Thimmayamma, B.V.S. 1981. *A Hand Book of Schedule and Guidelines in Socio Economic and Diet Survey*. Hyderabad, India: National Institute of Nutrition (ICMR).

UNICEF, FAO and SaciWATERs. 2013. *Water in India: Situation and Prospects*. New Delhi, India: UNICEF, FAO and SaciWATERs. Available at http://www.unicef.org/india/Final_Report.pdf.

10

Women's Voice in Water Resource Management

K.A.S. Mani, Vallaperla Paul Raja Rao, Madhukar Reddy and Ch. Ram Babu

10.1 Introduction

The Food and Agriculture Organisation (FAO) funded the Andhra Pradesh Farmer Managed Groundwater System (APFAMGS)[1] project. This successfully placed women in the forefront of water programmes and enabled them to assess the risks related to various water use practices and undertake group action (Rajeshwar Rao, 2006). This project prepared women not only as a pressure group but as equal partners by acquiring new knowledge. This strategy did not take the rural women through the feminist activist's path but prepared them for being central in decision-making process. Women were enabled to understand groundwater resource system, existing water use practices and level of misuse, and more importantly come out with clear recommendations on demand side groundwater management strategies for managing droughts. Thus, the project focussed not only on the women's concerns but also on environmental, climate change and sustainability issues.

Gupta (2009) notes that development status of women shows remarkable changes when drinking water is available, leading to their improved economic status. Ghosh (2007) exposed the commoditisation of natural resources and the specific ways in which it impacts women. Baruah (2007) examined the Self-Employed Women Association (SEWA) model in providing access to housing and financial services, especially low-income urban families living in slums. There are opportunities for extending this approach to other sectors. Literature on women's involvement in water is largely related to drinking water, sanitation and health, while more information is needed on how to integrate women into water projects, the need for supporting policies and successful case studies.

APFAMGS took several steps in mainstreaming women's role in groundwater management. The basic assumption was that the helplessness of women in times of distress was not always due to social and economic discrimination but more often due to lack of participation in the decision-making process and limited understanding. Increased women's participation combined with new skills, capacity and knowledge helped women to play active roles in the local institutions and thus contributed to the decision-making processes. APFAMGS Project Document (2003) had clearly identified a greater role for women in all stages of water resource management. Based on this understanding, gender equity was ensured in all stages of project implementation.

Haws (2006) noted that access to safe water and sanitation was important in removing the female face of poverty. Towards this end, APFAMGS ensured direct participation of women in different activities and ensured that changes taking place in the local community are significantly enhanced. More than participation, the leadership role for women was aimed at both the familial as well as the community level. Thus, it was ensured that women emerged as the face of APFAMGS project. Their participation in education, training, decision-making and governance was noteworthy.

Athukorala (1996) advanced the need for gender analysis in strategic planning for effective water management in Sri Lanka. Detailed gender analysis was taken up in the APFAMGS project for understanding the impact of water scarcity on women. APFAMGS Project Design saw the need to involve women in training, technology and decision-making so as improve their earning capacity, decision-making status, and more importantly, protect the status of girl children and thereby protect their freedom and health, and safeguard them against violence.

10.2 The APFAMGS Approach

APFAMGS project invested ₹300 million (over US$6 million) in the period 2004–2008. This was used largely for building capacity of the communities (50 per cent women) in data gathering and analysis, and empowering them with new knowledge related to the local groundwater system (see Box 10.1). Adequate support was extended from the local government agencies (₹138 million) through ongoing government programmes.

Box 10.1:
Participating in technical data collection through simple innovations

I am Ramadevi from Dharmapuram village, Miryalguda, Nalgonda. My family has six acres of farm land in which we have a borewell. Our crop production depends on the percentage of rainfall seeping into the borewell. In open irrigation wells, we can see the water but we cannot see water level in a borewell directly. Our borewell was selected for monitoring. Black pipes were lowered into them and water level indicators were given to farmers. My husband measured water level in our borewell twice a month and wrote it in Hydrological Monitoring Record. In the evenings, the farmers who had borewells sat on the village platform (Rachha Banda) and discussed water levels in their borewells. I heard my husband telling other farmers that he was able to understand groundwater resources better after he started measuring water level in his borewell. From that time onwards I too longed to measure the water levels because I spend most of my time farming. But, I could not do it because I did not attend school. I could not read the numbers on the tape and write them in the notebook. In one cultural programme (Kalajatha) it struck my mind that I can also use some tools I possess to measure water level. Thus, I started using the trick of using adhesive Bindi on my forehead. When the water level indicator made sound as it touches water I pasted adhesive sticker on the measuring tape and took home. With the help of children I recorded the reading. This procedure was also demonstrated at an observation borewell in presence of other GMC members to the women who had borewells. Thus, the dot of my forehead helped me to dispel the worry of illiteracy. I now understand the value of education. I am no longer helpless in making things happen.

This would help to implement farm-level activities related to increased water conservation, improved water use efficiency, crop diversification, organic farming and watershed activities. An additional ₹157 million was mobilised by the community on their own from various field-level interventions and inputs.

Despite this, IWMI (2012) cites the case of declining interest in villages (6 years after project closure) to sustain the programme where the institutions are not strong. Thus, it is very clear that NGO-supported institutions and the participation of women on their own cannot ensure sustainability of unconventional approaches. These programmes need active co-operation from the elected representatives, government funding and programmes.

APFAMGS worked with several government agencies to adopt the APFAMGS approach involving community especially women in data gathering and data analysis for sustained management of groundwater resources. The Ministry of Drinking Water (2012), Government of India, has adopted this model. It has implemented Rural Drinking Water Security in 15 blocks spread over 10 states specifically targeting women's group. These groups are encouraged to understand the reasons for existing drinking water crisis and take a lead role in ensuring drinking water security and improved sanitation. Andhra Pradesh Irrigation Department has scaled up the APFAMGS model to cover 13 districts as part of the Andhra Pradesh Community-Based Tank Management Project (2010). Ministry of Water Resources also plans similar models for implementation in six states. FAO has also initiated a follow-up project in the same area—Strategic Pilot on Adaptation to Climate Change or SPACC (2012) to organise farmers residing within a hydrological unit (HU, natural drainage basin). This is to build the adaptive capacity of the farmer to face the consequences of the drought, collectively. APFAMGS project was evaluated by various agencies to explore the scope for replication in the different states. FAO organised an evaluation at the project closure phase (2008). World Bank independently carried out its own evaluation and brought out its finding as a special Publication Deep Well and Prudence (2010). The Institute of Rural Management, Anand (IRMA) students also carried out an evaluation of APFAMGS project to understand the community role in implementing the programme six years after closure with similar findings as IWMI (2012).

10.3 Addressing Poor Representation of Women

Local institutions in India have generally been used by men to derive political and personal gains, while women have been included only to meet constitutional obligations. Women seldom participate in the decision-making process. This approach was challenged for creating the enabling environment for women to use local institutions to affirm their rights and make their voice heard. Village women were for the first time taken on international visits to understand the conditions and assess opportunities for group action. APFAMGS has systematically documented the new actions of rural women for improving their understanding of the local groundwater system and groundwater availability. A special publication 'Trail of Change' (2006) has brought out several such stories.

Village-level Groundwater Management Committee (GMC) and federation of villages under Hydrological Unit Network (HUN) that was established under the project guaranteed 50 per cent membership to women. More importantly, all decision-making resolutions stipulated 50 per cent attendance of women (quorum). Women's active role in local institutions resulted in recognition of women beyond an appendage to man and the family. Special trainings were given to men in all the 650 villages in the project area to understand women's concerns and the new opportunities for addressing water distress collectively as equal partners. Active roles for women ensured male members overcame selfish interests, inculcated new value systems and worked towards common goal that benefited everyone.

Active participation of women in local institutions (Table 10.1) has helped ensure gender justice. It enabled women to have an equal role in technical data gathering, data analysis, decision-making on crop water use, use of chemical fertilisers and pesticides, marketing of agriculture produce and more importantly having a say in the nature of investments to be made. Institutional approach to groundwater management not only got the recognition of women as important stakeholders but also identified the social grounds access to information and right to decision-making that make them more vulnerable to any crisis related to shortages. Improved groundwater governance cannot happen in isolation. It must be worked along with gender-related issues such as preferences, inequalities in access to and control, access to technology and more importantly a critical role in the decision-making process related to groundwater management.

10.4. Bringing Technical Data Gathering into the Public Domain

Scientists design hydrological data gathering for universal use. However, the technical specifications make it complicated and difficult for the ordinary water user to appreciate its relevance. The data gathering is always

Table 10.1:
Membership of women in decision-making bodies

Name of NGO	General Body Members of Village Groundwater Committee		Executive Body Members of Village Groundwater Committee		General Body Members of Village Federation Groundwater Committee		Executive Body Members of Village Federation Groundwater Committee		Percentage of Women in Different Committees	
	Women	Total	Women	Total	Women	Total	Women	Total	Village Committee	Village Federation
BIRDS	295	706	119	343	80	165	18	55	42	48
CARE	213	581	169	383	59	118	10	29	37	50
CARVE	254	632	159	403	76	167	47	109	40	46
DIPA	320	746	410	574	165	247	35	49	43	50
GVS	174	443	214	279	70	108	36	47	39	54
PARTNER	192	514	280	392	112	169	31	44	37	51
SAFE	219	549	350	483	160	235	49	67	40	47
SAID	227	512	270	374	104	156	49	70	44	50
SYA	174	437	437	611	118	177	71	101	40	50
Total	2,068	5,120	2,408	3,842	944	1,542	346	571	40	50

done by government agencies, research institutes, academic institutes and others specialised in technical skills. The first task in front of the project was to de-mystify the process and make it socio-technical so as to encourage participation of all. Data gathering was introduced in the rural habitations adopting very strict technical guidelines, yet making the data available real time to the users. Thus, the monopoly of access to scientific knowledge was broken, but all the more important was that it was made equally available to women in APWELL Project (2003). Limited technical knowledge has been held as an excuse for not involving women in the decision-making process. This was broken by the project by ensuring that male and female members from the same family work as a couple in data gathering, analysis and interpretation. Trainings were offered to couples. Data and knowledge were promoted as an opportunity for bringing gender balance, exposing women to acquire unconventional skills, handling different technologies, and more importantly, accessing associated knowledge in order to participate in the development process as equals (see Box 10.2). Community data gathering has been a major achievement and data (Data Products Catalogue, 2008) was purchased by various national and international groups.

Box 10.2:
Understanding the reasons behind groundwater depletion

Subharathi a woman farmer, living in Bachupalli, Gooty mandal of Anantapur district, Andhra Pradesh, India, is one of the few women members of the village groundwater management committee. She says, The groundwater management committee was actually measuring the groundwater levels, quantity of water pumped from the wells, rainfall and water quality. I was so excited. This was interesting and so different to the things that I do at home and in the farm. I looked forward to the days when we would meet to discuss about the groundwater problems. This was the only occasion where I had the opportunity to be myself and give views on all issues. In fact, I could sit alongside other male farmers. This is a new social status. I did not want to let go of this opportunity as it gave me equal status that I always missed but could never express. When the village water supply was badly affected, the groundwater management committee decided to approach the rural water supply department and we successfully got it repaired. This gave me the confidence to believe that we could solve all our problems by coming together and approaching the concerned department. After regularly collecting water level, pumping data and rainfall data for a year, all the groundwater management committee members carried our data and sat along with farmers from neighbouring villages of Upparavanka drainage. Only in this meeting, I came to know that all the habitations belong to a single drainage family, so others who were there are my brothers, sisters and cousins. Those who are in the upper side of our habitation are the elder brothers/sisters and the ones below are the younger brother/sisters. We all have to respect each other and follow the family tradition of sharing and giving. This suddenly opened my eyes and left me ashamed about my ignorance. In this meeting, all the data brought by the various habitations were entered into a single book as a joint family account and earnings/spending worked out. Many farmers who have been members for long were participating enthusiastically. Seeing them I said to myself, next year I will be in a position to take part as actively as them. This is a challenge I have set to myself. After entire accounts were worked out, it was told that expenditure exceeds income. The groundwater pumping exceeds the availability and hence the failure of borewells or falling groundwater levels, and it was no curse of nature. This I believe is like withdrawing more money from the bank account than I have saved. This means clear indebtedness. I was always taught not to take loans. I always managed my expenditure matching with our income. Now here I am in a drainage, tapping more groundwater than I possess. I feel ashamed. Now I am working to ensure that my fellow farmers and I draw only that much amount of water from wells that we are eligible to. I do not believe in the virtue of overspending.

To ensure gender justice, it is important that both men and women have equal access and control over natural resources, technology, local institutions and constitutional safeguards. Adopting this approach, the project encouraged women to participate in gathering water-related data and thereby get better understanding of the resources. Both men and women were trained in data collection and analysis, and understood changes taking place in the groundwater resource availability. Community members measured daily rainfall, bi-weekly groundwater levels, groundwater discharge, daily surface runoff, seasonal water quality and seasonal crop details. Data gathered were made available on real-time basis to the entire community using village display boards. All data were discussed in the local institutions, analysed and interpreted for taking decision related to its use.

10.5 Farmer Water School

Adopting the theory of 'seeing is believing' and the concept of peer group learning, Farmer Water School (FWS, 2007) methodology was adopted for training and capacity building of men and women. The uniqueness of FWS is that it looks at training as a process towards empowering the learners with knowledge and skills to understand the total water system. The trained FWS graduates get a certificate on successful completion. They not only get to learn about the water resource system but also understand the implications of all of their actions on overall resource availability. Additionally, this approach helped sensitise both men and women on the need for collective action for effective management of the resource within the concerned hydrologic unit. FWS helped farmers' ability to make critical and informed decisions on crop plans and water use efficiency so as to get the best economic returns.

Access to scientific data helped women to understand the system, the nature of the crisis, as well as linked economy to the resource. New knowledge and a clear role in the local institution helped provide women a new voice to exercise their rights in critical decision-making processes related to water use, its management and conservation in drinking water as well as irrigation sector. Three pillars of success—women's participation, demystification of science of water and improved local governance—helped design farm and village-specific water use practices, which helped manage the water crisis that resonated appropriately at a regional hydrological unit level. Empowerment helped bring out clearly in the open connections between the right to water and gender justice (GAS, 2005).

10.6 Transforming Data to Knowledge

Data gathering by community makes sense only when they acquire new information, based on which they can formulate new hypotheses about the local water crisis. The whole process of sorting information from the data has been accomplished using the platform of crop water budgeting (CWB) workshop. CWB is an annual event conducted in the months of

September and October prior to the Rabi crop (groundwater-based cropping season). In the workshop, men and women from different villages yet part of the same drainage family come together. They collated the data gathered for computing the total water balance and understanding the available recharge as against the total groundwater pumped. The demystification of technical data into simple assertive information has had far-reaching consequences within the community.

Thus, the disconnect between reality and apprehension, on the one hand, and the wide gap between water availability and its sustainable use, on the other, is clearly understood. The community is challenged to undertake appropriate steps and even take tough decisions to confront the problem head-on and bring about positive transformation in the life of the most vulnerable communities. Women for the first time get to access critical information. As a consequence, they express views on the problem and suggest solutions, which are more environment-friendly. They also focus on home needs (food, nutrition, sanitation, health), which are less commercial and risky and more importantly do not ask for large quantities of water. It is for the first time that rural women carried out inventory of local greens, fruits, grass and herbs that contribute to local food and nutrition. Their findings have been documented as a publication 'Nourishing Traditions' (APFAMGS, 2006). This was recognised by Nutritional Society of India as new information that also needed appropriate follow-up for protection and revival. Some of the solutions offered are highly innovative and if implemented will be most beneficial in preserving and protecting local herbs, fruits and medicines while at the same time make available high-value food and nutrition at very low cost.

New understanding on the water crisis helped the community to think and act together. It minimised risks associated with water stress and maximised the benefits from the available water. Field level actions included increasing the crop varieties from 18 to 45 over five years. The range of crops helped reduce pest attacks, spread soil moisture availability over long periods, reduce risks of crop failures, and minimise investments on chemical and inorganic inputs. Turning to organic farming reduced the water demand and the cost of crop production, ensuring profits. Adoption of vermi-compost, mulch and soil alteration increased the soil moisture levels over longer time periods while reducing crop water demand. Implementation of drip, sprinkler, water guns and other tools resulted in reduced water demand from as much as 40–60 per cent. Combination of various techniques helped reduce groundwater pumping in over 200 project villages in three years. In other villages, the impact is of varying order and the time to see positive results will be spread over a longer period of time (see Box 10.3).

Box 10.3:
Implementing crop diversification as a means for saving water

I am Pidathala Guravamma of Salakalaveedu in Besthavaripeta mandal of Prakasam district. This time, I did not heed to my husband's advice on choice of crops. I followed my own ideas and my crops are growing well. In the year 2003, CARVE-APFAMGS came to our village and conducted meetings in the temple. They conducted a Kalajatha programme and selected some borewells for observation, and installed a rain-gauge station. But the women of our village did not come out to participate in the programme because their men didn't allow them. The staff from CARVE-APFAMGS began to motivate the women to participate in the programme. That is how I joined the Groundwater Monitoring Committee (GMC) as a member. Initially my husband did not like, but as I explained the project goal and objectives, he understood. I attended several training sessions on water level measurements, alternative agriculture, irrigation methods, role and responsibilities of GMC members, bookkeeping, etc. We conduct an annual workshop on crop water budgeting as a part of farmer water school. I attended all the sessions and learned many things. I had a chance to attend another workshop in Khambam on crop water budgeting. In this workshop, we learned about groundwater situation of Pulivagu HU. They taught us about water budgeting, drip irrigation and rice cultivation by System of Rice Intensification (SRI) methods. They also mentioned the shortage of groundwater in Pulivagu HU. The workshops ignited many thoughts in my mind. I realised that we were committing a mistake by wasting lot of water without knowing its value. I decided to stop this immediately by changing crops. We wanted to grow cotton, chilly and sweet orange but the influence of Crop Water Budgeting workshop was so strong on me that I planned to grow groundnut and to adopt drip irrigation. I shared my views with my husband and he agreed. I shared my experience with farmers of 11 villages who attended a 'feedback' programme on Crop Water Budgeting workshop, conducted on 22nd of December. All the farmers appreciated me. I felt very happy and proud at that moment. I thank the local support with all my heart for changing me and giving me the confidence to share my experience with fellow farmers.

10.7 Women's Role in Water Governance

The APFAMGS approach is an attempt to create an autonomous space for women to articulate and demonstrate their strengths, vision and concerns in the management of sustainable development of groundwater. The methodology attempted is ground-breaking. It moves away from the old-fashioned approach of token participation without accommodating women's views. Critical in this approach is women's leadership role in the local institutions; taking up technical data gathering; participate in estimation of the resource availability; assessing the nature of the crisis; developing consensus towards taking appropriate solutions and then enforcing discipline through local governance.

The critical issues that were considered as part of local governance include the following:

- Treating groundwater as a common property resource with local institutions laying down the general norms for sustainable use.
- Moderate and curtail the unhealthy competitive drilling of bore-wells by water users that affect the neighbourhood hydrologic system.
- Develop protocols of water efficient crop plans that match with the water resource availability in the local areas.
- Prioritise drinking water needs and food crop needs while developing water use plans.
- Ensure adequate production of required food and nutrition crops and thereafter only plan for cash crops.
- Ensure organic crop production as much as possible in order to prevent contamination of the local groundwater system.
- Adopt water-saving technologies using sprinkler and drip, conveyance of water through pipelines, and improving soil moisture availability using vermi-compost, tank silt, mulch, plant residues and soil treatment.

Many of these rules were proposed by women themselves and supported by the entire committee. The approach has offered a very new experience for the women to participate as equals, express their own views without compromising on their needs, and more importantly demonstrate and show to the community their resolve to accomplish the tasks.

The overall gains from active women's participation have had a sobering influence on the attitude and value system of male-dominated society. The project could decisively prioritise on issues related to food security, drinking water, health, nutrition, sanitation, environment, sustainability, water use efficiency and economic returns from a water drop. This has happened for the first time in any irrigation project in the country without in any way altering the work plan or increased spending.

10.8 Conclusion

The APFAMGS project, recognising the vulnerability of women to gender-based discrimination, especially in time of water distress, taps into the innate qualities of women, that is, their intelligence, capacity, skills and knowledge for the common good of the community. This has helped to make radical changes to the whole concept of water management and water use. The project has also partnered with youth and used their services to address their concerns on drinking water, irrigation water, agriculture, sanitation, food, nutrition and health.

Adequate investments in capacity building of women help village-level decisions related to prioritisation of water use efficiency, linkages of food and nutrition with crop plans, and ownership and access to water. The APFAMGS project has clearly demonstrated that men alone cannot handle water distress. Further, it has shown that equal partnership of women is a basic necessity and all government programmes need to recognise and involve women in the project design and implementation.

APFAMGS model has received several recognitions, including global recognitions (shortlisted as 10 best global water project, Kyoto Grand Prize, IV World Water Summit, Mexico, 2006). Its approach is seen to be relevant in South Asia, Africa, Middle East and South America (where gender discrimination is prevalent), where water crisis such as droughts as well as floods, seawater ingress, natural contamination and industrial pollution makes it increasingly relevant. The Government of

India has also recognised the relevance of APFAMGS approach to the entire country and has incorporated it in the Approach paper to the XII Plan Document (2012).

Experiences from around the world have shown that woman's concerns when prioritised, automatically addresses the sustainability issues. Appropriate frameworks, legislations and policies still need to be developed to ensure women's role in local-level data gathering, data analysis and local governance with an overarching water law that acknowledges women's concerns that are consistent with the national framework.

It is fair to conclude that initial challenges for involving women in natural resource management are huge, largely because of their lack of confidence and negative male attitude towards female capabilities. However, once women get past their inhibitions, their ability to make informed choices is revealing. Their ability to look at larger issues is phenomenal and they can pull together such diverse strands as food production, nutrition, environment, organic farming, drinking water security, reuse of waste water and water quality protection.

This experience has shown that now is the appropriate time for water resource system planners to create new social and gender norms in all aspects related to natural resource use and its management, both for their own empowerment and for sustainability.

Note

1. Andhra Pradesh Farmer Managed Groundwater System (APFAMGS) project implemented in Anantapur, Chittoor, Kadapa, Kurnool, Mahaboobnagar, Nalgonda and Prakasam districts of Andhra Pradesh introduced Participatory Hydrologic Monitoring (PHM); http://www.fao.org/nr/water/apfarms/index.htm.

References

APFAMGS. 2006. Publication No APFM/RE/60/2006 'Nourishing Traditions' Local Greens. http://www.fao.org/nr/water/apfarms/upload/pdf/RE6006_40125.pdf

APFAMGS. 2008. Publication Data products Catalogue. http://www.fao.org/nr/water/apfarms/upload/PDF/Data Products Catalogue.pdf

APFAMGS Project Document. 2003. http://www.fao.org/nr/water/apfarms/Project_Document.htm

APFAMGS Publication. 2006. No APFM/RE/50/2006 "TRAIL OF CHANGE" villagers view point on APFAMGS intervention-Vol I. http://www.fao.org/nr/water/apfarms/upload/pdf/RE502006_40035.pdf

APWELL. 2003. Participatory Hydrological Monitoring. http://www.sswm.info/sites/default/files/reference_attachments/APWELL PROJECT Participatory Hydrological Monitoring Manual.pdf

Athukorala, K. 1996. The need for gender analysis in strategic planning for effective water management in Sri Lanka. *International Journal of Water Resources Development*, 12(4): 447–460.

Farmer Water School (FWS). 2007. Concept adopts the technique of demystifying the science of water and empower water users with knowledge and skills to take appropriate steps at the farm level and local drainage unit level to manage the water crisis. http://www.fao.org/nr/water/apfarms/upload/pdf/RE7907_40108.pdf

Baruah, B. 2007. Assessment of public–private–NGO partnerships: Water and sanitation services in slums. *Natural Resources Forum*, 31(3): 226–237.

Gender Analysis Studies (GAS). 2005. Were carried out as part of baseline data gathering in parts of 7 districts of Andhra Pradesh spread over 638 villages to understand the role of women in water management. http://www.fao.org/nr/water/apfarms/Gender_In_Water.htm

Ghosh, N. 2007. Women and the politics of water: An introduction. *International Feminist Journal of Politics*, 9(4): 443–454.

Gupta, D. 2009. Disparities in development, status of women and social opportunities: Indian experience. *Journal of Alternative Perspectives in the Social Sciences*, 1(3): 687–721.

Haws, N.J. 2006. Access to safe water and sanitation: The first step in removing the female face of poverty. *Women's Policy Journal of Harvard*, 3: 41–46.

IWMI-TATA Water Policy Program Publication no 37. 2012. Andhra Pradesh Farmer managed Groundwater System Project (APFAMGS) a reality check. http://www.iwmi.cgiar.org/iwmi-tata/PDFs/2012_Highlight-37.pdf

Ministry of Drinking Water. 2012. Note on Water Security Pilot. http://www.mdws.gov.in/node/2406

Note on Andhra Pradesh Community Based Tank Management Project (APCBTMP). 2010. http://www.apwaterreforms.in/apcbtmp.html

Note on Strategic Pilot on Adaptation to Climate Change. 2012. http://coin.fao.org/cms/world/india/en/Projects.html; http://www.fao.org/nr/water/apfarms/upload/PDF/GCPIND175NET-APFAMGS-eva-final.pdf

Rajeshwar Rao, Salome. 2006. Technology and Knowledge for Gender Equity and Justice. APFAMGS Publication no APFM/CP/53/2006. http://www.fao.org/nr/water/apfarms/upload/PDF/CP532006_40031.pdf

World Bank Publication. 2010. Deep Wells and Prudence Towards Pragmatic Action for Addressing Groundwater Over-exploitation in India. http://siteresources.worldbank.org/INDIAEXTN/Resources/295583-1268190137195/DeepWellsGroundWaterMarch2010.pdf

11

Leadership and Participation: Role of Gender

Sudhir Prasad, Satyabrata Acharya and Somnath Basu

11.1 Introduction

The state of Jharkhand faces a paradox of having huge water shortage in the dry seasons despite generous annual monsoons (annual average 1,600 mm rainfall) every year. The water shortage in the dry seasons has serious implications for the drinking water sector, as the rural population resorts to unimproved sources[1] to meet their demand for drinking water. The census data indicate that 44 per cent (Census, 2011) of the population living in rural Jharkhand are still accessing drinking water from unimproved sources. It also indicates that only about 4 per cent of the rural population have house connections for drinking water through piped water supply systems. The Drinking Water and Sanitation Department (DWSD) of the Government of Jharkhand has decided to reverse this scenario by increasing house connections through piped water supply to 25 per cent by 2017. This would reduce the percentage of the population accessing unimproved sources for drinking water (Mitra, 2011).

While the DWSD has taken initiatives to increase the coverage of drinking water by piped water supply and home connections, impending issues of ensuring sustainability of the systems and equitable access remain. Hence, the DWSD, as a matter of policy has adopted a demand-driven approach with intensive community involvement.

Against this backdrop, the implementation of drinking water supply systems in *Belkhera* village in Koderma district of Jharkhand state serves as a significant aspiration for the state government. This project implemented by PRADAN as part of corporate social responsibility (CSR) for the Damodar Valley Corporation (DVC), provided a model for the promotion of piped water scheme with the involvement of self-help groups

(SHGs). The SHGs were involved in planning and implementation of this scheme. They also managed the scheme by charging water tariff from the users. Interestingly, all the 75 households in the village have water connections. The *Belkhera* experience depicted that the role of women in the SHGs is very significant as women unequivocally play the central role in the sustainability of the scheme.

What would then constitute a successful community-based water supply programme in India? What role should the government play in setting up such a community-based system? Which grassroots institutions can support community-based systems most effectively? Equitable access to drinking water is also a key concern that needs to be addressed; and in that context, gender equity is of critical importance. To bring the gender issues of participation in implementation of the water supply schemes and ensure that their voices are heard within the framework of a participatory process are key challenges of the sector.

As the DWSD is gearing up to establish a policy for implementing sustainable and equitable drinking water supply systems at scale, these questions need to be addressed. It was felt necessary to develop a standard set of guiding principles that will ensure that community-based water supply schemes are reproduced in a routine manner, irrespective of socio-economic backdrop of the state. A pilot of community-based water supply programmes in Jharkhand has been initiated in 20 different villages spread across 9 districts with the objective of developing the standard operating procedure for *drinking water supply projects with community participation and using government's funds*. This we suggest can be universally applicable.

This chapter contains five sections. The first section presents the overall perspective of the DWSD and the background to the piloting of the community-based drinking water supply schemes. It focuses on the role of women groups in planning and implementation. The second section explains how the different grassroots institutions that are involved in the implementation of the schemes, their composition and inter-relationship. The third section elucidates the study design and specifies the geographical locations where the field research was conducted. The fourth section presents the findings of the study. The socio-economic backdrop in which the different grassroots are functioning is described. The behavioural pattern of communities belonging to different population groups have been studied and categorised. Three critical factors that characterise the involvement of women in the drinking water sector are then deciphered. These factors are (1) women in leadership, (2) women in participation and (3) user's perspective of the women or consumers of the water supply schemes. In the fifth section, the findings of the study have been summarised along with key recommendations to support policy formulation of the drinking water sector in Jharkhand.

11.2 The Pilot Project on Community-based Water Supply Programme

Based on the *Belkhera* model, the pilot community-based water supply programme for 20 villages was implemented through a tripartite partnership. This was between the DWSD under the Government of Jharkhand, PRADAN—an NGO facilitating the entire field level implementation, and UNICEF providing support in field level coordination, monitoring and reporting. Funds available through the National Rural Drinking Water Programme (NRDWP) is used for hardware installations and the Village Water Sanitation Committee (VWSC), a statutory body set up under the provisions of the *Panchayati Raj* Act, is the executing agency, that is, recipient of the programme funds. The elected *Mukhiya* (elected head of the local self government at the *Gram Panchayat*) of the *Gram Panchayat* is the President of the VWSC. The position of Vice President is reserved for women and she has to be an elected member of the block or district *Panchayat*; this is done to ensure representation of all three layers of the *Panchayats* in the VWSC.

The villages/communities where PRADAN's livelihood interventions through SHGs were strongly established were selected for the pilot interventions. It may be noted that 50 per cent of the members of the VWSCs are women, as per the statutory regulations. In the pilot villages, the women members of VWSC very often coincided with women members of the SHGs supported by PRADAN. Given that all the six or even if a few women members of the SHGs were present in the VWSCs, the entire SHG would rally around the VWSC and provided enormous support at the grassroots for consensual decision-making based on the existing community bondage established through the SHG activity.

A process of inclusive planning has been followed to ensure equitable access to the water supply scheme. Thus, all families and institutions in the villages will have house connections. Accordingly, the detailed project report (DPR) has been prepared with PRADAN's technical assistance and submitted to the DWSD. With the funds transferred till date, the VWSCs have created the water source, that is, seepage wells. The construction of the remaining part of the scheme, such as the pump house, overhead tanks, lying of pipelines, etc., is under progress.

11.3 Study Design and Methodology

11.3.1 Study Design

This study is based on qualitative analysis of information and data obtained from the field. Data have been collected through Focussed Group Discussion (FGD) conducted with men and women. The findings of the current study are based on qualitative research conducted in the project sites. The study focuses on 12 villages in five districts of Jharkhand. The districts were selected to represent three different geographical areas of the state, namely, North Chotanagpur (Hazaribag and Koderma), Santhal Parganas (Dumka) and Kolhan (Khunti and West Singhbhum) (see Table 11.1). Thereafter, six villages, distributed in the three regions were selected for the field study. Since these villages were in an advanced stage of project implementation, participation and leadership issues were well addressed here.

Table 11.1:
Distribution of study villages for focussed group discussion

Region	*Districts*	*Villages*
Santhal Parganas	Dumka	Dharampur, Darbarpur
Kolhan	Khunti	Rohne
	West Singhbhum	Kendposi
North Chotanagpur	Koderma	Jolakarma
	Hazaribag	Petula

Intensive FGDs have been conducted in six villages (mentioned in Table 11.1) with structured questionnaires. In each village, FGDs were conducted separately with VWSCs and SHGs (all women) and men. At an average, 30 individuals participated in the process in each of the six villages, thus involving approximately 120 individuals for the study. In another eight villages (four in North Chotanagpur region and two each from Santhal Parganas and Kolhan) observations and discussions were held with unstructured questionnaires. In these villages, the progress of work was relatively slow as the community had less experience in dealing with real issues of programme implementation. However, the information generated was useful to supplement the comprehensive findings from the former six villages. Discussions were also held with the field level executives of PRADAN to develop an understanding of the functioning and group dynamics of SHGs.

11.3.2 Key Research Areas

Three different aspects of the role of women have been studied as mentioned earlier. Firstly, the leadership role of the women has been studied. Structured FGDs were conducted with different village level groups to understand how and under what circumstances women emerged as natural leaders in leading water supply schemes. Secondly, women's participation was studied with a view to understand the barriers to participate in certain activities and how these barriers are overcome. Thirdly, the role of woman vis-à-vis consumption of drinking water at the household level has been studied.

11.4 Findings

11.4.1 Setting the Socio-economic Backdrop to the Study

At the outset, it is necessary to illustrate the socio-economic backdrop in which the community processes were studied. Jharkhand is primarily a tribal state. Total 26.3 per cent (Census, 2011) of the population is tribal, though the tribal population is not homogeneous (Sharma, 2004). There are different sub-groups among tribal population,[2] and they distinctly different from each other in social norms and community processes; the only common denominator being the landscape in which they live—predominantly forest land (28.1 per cent; Census, 2011) and very poor social infrastructure (Prakash, 1999, 2007).

In Table 11.2, we map the characteristics of the population at the micro-level with particular reference to community participation vis-à-vis SHG movements. These characteristics have also influenced the working of the VWSCs and the SHGs.

The following observations are based on the field study to capture individual and community behaviour.

In Santhal Parganas for instance, the SHGs and the VWSCs are subsumed within each other; there is no conflict between the groups and

Table 11.2:
Regional variations in social and economic features

Region	*District*	*People*	*Economy*	*SHGs*
Santhal Parganas	Dumka	Santhal tribes: extremely submissive and introvert. Not forthcoming in front of external agencies; leadership is generally collective in nature	Extremely poor, the area has poor infrastructure but is rich in mineral/ coal reserves; historically the tribal are marginalised	Moderate: They have responded to the SHG initiatives to some extent and ensured basic livelihood security, though no tangible improvement in quality of life is visible
Kolhan	Khunti, Chaibasa	Munda, Oraon: not very enterprising on their own but can follow up very effectively when directed; both individual and collective leadership visible	Moderate economic status, huge public investments in development of social infrastructure; mineral/coal rich; tribal are historically exploited	More proactive compared to the Santhal villages; participation is more spontaneous and livelihood security is followed by sporadic improvement in standard of living

(Table 11.2 Continued)

(Table 11.2 Continued)

Region	District	People	Economy	SHGs
North Chotanagpur	Hazaribag, Koderma	Caste Hindus (upper castes, OBCs and SCs); mostly dominated by upper castes, even if representation from lower caste are taking place the community dynamics are dominated by the upper castes	Basically an extension of Bihar, populated by caste Hindus; have typical social discriminations, comparatively richer; most of the income through remittances from a huge migrant male population	Prosperous: Here the women have actually seized the opportunity provided by the SHG platform. They have generated sizeable income from the economic activities (farming, poultry, etc.), saved judiciously created household assets to support further increase in income

Source: FGDs held by the authors (these perceptions are based on field experience; while implementing the pilot project with SHG initiatives extensive FGDs were held by the authors).

both the groups have identified water as their common goal. Individuals in the Santhal villages do not compete for leadership. Collective leadership is prevalent as groups of women work together with a common cause. Women, by and large, are not subjected to discrimination in decision-making and participation. In fact, in most social forums, they represent households and communities. Hence, the collective representation of women is basically the manifestation of leadership in these areas.

In Kolhan area, the community is comparatively more enterprising, due to comparatively better social infrastructure. As in case of Santhal Parganas, the tribal society is inclusive in nature. Thus, both VWSCs and SHGs have exhibited a tendency to work with each other. Though the community in general is comparatively more proactive than the Santhals, collective representation is the overarching principle of community dynamics. Often there is resistance to and conflict with external agencies such as major political parties or government officials whom they have to deal with on a day-to-day basis for scheme implementation.

The third significant variation exhibited by the groups of the North Chotanagpur area. The population in this region is largely non-tribal.[3] There are various layers of social discriminations characterised by caste, class, religion and gender that work simultaneously making the social fabric very complex. The gender alignment prompted by the SHG movement in these areas is a remarkable step towards ensuring gender parity, apart from improvement in livelihood status. The SHG movement vis-à-vis women's empowerment has also offset many other social disparities and has empowered the voice of the vulnerable. Here, the VWSCs constituted under the provisions of the *Panchayati Raj* are in several cases influenced by the socially and economically dominant groups and mostly represented male members.

Another critical context of this study is the interaction between the VWSC and the SHG. The VWSCs are the statutory bodies formed under the Indian constitution. The SHGs are a consolidated group of 20–25 women brought together through a process of collective realisation of community and family benefits. The range of issues that the SHGs deal with start from livelihood issues at the household level to basic amenities at the community level. The impending question is how a constitutionally defined and delineated group interacts with a dynamic evolving women's group in a socio-political space and influence each other in decision-making and other operational aspects. The issue of *women and water* needs to be viewed in this backdrop as well.

In the following sections, the key findings of the current chapter are based on the above backdrop.

11.4.2 Leadership

The network of SHGs, facilitated by PRADAN, triggers community participation in the water supply schemes in all the cases. Also, the recognition of the VWSCs as a constitutional body eligible for receipt of government funds has evoked a sense of responsibility and ownership. A combination of these two factors has led to community processes wherein the SHGs strengthen the community processes, while the VWSCs provide the constitutional strength to the community actions.

Under such circumstances, how is the issue of leadership construed? On one hand, the issue of 'leading from the front' may be viewed as leadership; on the other hand, building consensus in a group collectively without emerging as an individual spearhead may also be viewed as leadership. It has been observed that both in the Santhal Parganas and Kolhan region, women within the community do not contest with each other for individual leadership. The society is characterised by gender parity, and women generally represent families and communities in social forums. The collective views of the women are heeded in groups either in line with the constitutional framework or socio-economic issues.

The traditional value-based role of the village head (*Manki Munda*) has also helped in establishing social equity and traditional gender parity. The traditional social values guide the behaviour and aspirations of both men and women. During the FGDs, it was been noted that individuals do not express themselves readily. However, they are aware about their roles in execution of the drinking water project. Leadership is viewed not necessarily as a position of extraordinary privilege. They do not assign exceptional significance to one or two members from the group who would tend to express their viewpoints; once an opinion is expressed, most often the remaining members of the groups accept them. For a collective cause, the women generate consensus without much conflict.

In Kolhan, there are instances of one or two proactive women members who stand out in the group, but without carrying any extra *privilege*. Women's leadership in this area are driven by a desire to serve the community and override the desire to seek special status in the society (see Box 11.1).

Box 11.1:
Illustration of selfless leadership in tribal communities

The case of selfless leadership is best exemplified in case of Kendposi village of Chaibasa. Here, interestingly, Baijmati Sinku emerged as a women leader by virtue of the fact that she is the President of the Block *Panchayat*, the Vice President of Kendposi VWSC and also a seasoned SHG member. She used her position to counter a range of difficulties that the Kendposi water supply scheme was facing due to a politically motivated *Mukhiya* (President of the *Gram Panchayat* and also President of the VWSC). Evidently, the process of empowerment started with her association with the SHG and further strengthened through her election to the Block *Panchayat*. In the village (which has 100 per cent ST population) she has full confidence of the people—most interestingly without any visible opposition. The acceptance of Baijmati Sinku in the community may be attributed more to the social norm of equality; also she does not seek exceptional privilege for her leadership role.

However, in North Chotanagpur region, any positions for which individuals are nominated, elected or selected are viewed as privileged entitlements. Hence, both the formation and the functioning of groups are much more complex here. The group domains are clearly demarcated and need the facilitation of an external agent or agency for functioning. This situation is avoided if more than 50 per cent members represent both groups, that is, VWSC and SHG. Most significantly, these populations have a tendency to have women as token representatives in village groups. The male counterparts represent the women in all group activities. At times, husbands and fathers fulfil the assigned responsibilities compromising the aspirations of women.

In the given social fabric characterised by different layers of discrimination, the community usually accepts this arrangement. However, a substantial presence of the SHG members in the VWSC could resist family interests and influences in the working of the VWSCs (see Box 11.2). It was also noted that simply by ensuring the involvement of SHGs, a range of discriminations is addressed. In the SHGs, the women align and protect each other's interests—which may be either at the family level or at the community level. It has been observed that despite dominance of the socially and economically influential people, women as a group could

Box 11.2:
SHGs ensure inclusive development and gender parity

In Jolakarma (Koderma), a male member, an opinion leader, Azim Ansari was the key driver of the water supply scheme in the village. He has substantial credibility in the community, earned through years of social services. His interest in the VWSC can be derived from the fact that his wife is the *Upadaksh* (Vice President) of the VWSC. Though some SHG members are incorporated in the VWSC, they have limited influence in the committee. Thus, though Azim Ansari's enterprise in implementation of the water supply scheme is laudable, various underlying disparities propelled by caste, class and gender remain compromised. This evidently results in exclusion of certain sections of the population; also it will most certainly have an impact on the sustainability of the programme. The women SHG members here act as a '*regulator*' of the VWSC, contesting various actions such as (a) process of formation of VWSC, (b) manner in which VWSC conducts its business and (c) manner in which benefits are to be shared.

express and establish their viewpoints. While dominance of the men cannot be ignored, women have nevertheless, created a space for themselves and were vocal during the deliberations.

11.4.3 Participation

The role of women in the physical execution of the project is a critical measure of their association and ownership of the project. This has been observed in case of the *Belkhera* model. However, the dimension of women's association in the six different project sites was found to be varying in magnitude and character. As indicated before, the overlap between VWSC and SHGs resulted in better representation of the women. This overlap is more visible in Santhal Parganas and Kolhan where the society is more or less homogeneous (tribal population). There have been instances of individual office bearers who are politically motivated or have personal vested interests within the VWSC. However, the members of SHGs in the VWSC have ensured universal participation and that consensual decisions are respected.

Again, the level of women's participation in the Santhal Parganas and Kolhan has been more pronounced. In these areas, women have been instrumental in two key aspects, namely (1) interaction with the government departments and (2) supervision of civil works. Women, in most cases, have significantly represented the VWSC. Even if the *Mukhiya* was a male member, group of women would accompany him to the office of the Executive Engineer for submitting the DPRs or to collect the cheques. The diligence of the women who took lead in these matters was remarkable, as these functions require multiple visits to government offices or the banks. The women have been steadfast in their resolve and ensured that the first stage of the project execution, that is, construction of well is properly completed.

The level of participation of the women in physical execution of the project varied from the conventional. These ranged from construction labour to the unconventional, levelling with theodolite. They were more intensively involved in supervision, procurement of materials and maintaining the book of accounts. Evidently, this intensity of involvement was triggered by their perceived benefits due to availability of drinking water at home. Also, it was possible due to their empowerment in the process of being associated with the SHGs. The above picture was palpable in all sites of Santhal Parganas and Kolhan (see Box 11.3).

Box 11.3:
Women bolsters local self governments; more so in tribal communities

In Ronhe village of Khunti district, the President of the VWSC, Immanuel Munda, a tribal man, was a typical example of a motivated *Panchayat Mukhiya*. He is fully dedicated to the water supply programme. Immanuel Munda has limited exposure to the broader domain of politics and administration and hence his interests are confined to the villages from where he has been elected. The social space of the village is dominated by the SHGs and the *Mukhiya* is also under the influence of the progressive vision of the village women. The result of this interface has been exemplary. The VWSC and SHG has become co-terminus. Here, the SHG women have become a huge strength to the VWSC in general and *Mukhiya* in particular.

The level of participation is different in North Chotanagpur region. Here, the social fabric determined the level of overlap between the VWSC and the SHGs and subsequently the level of participation of the women. The FGDs indicated that in some areas, the SHG and VWSC are not in sync although the women have identified water supply as a critical issue in their lives and expressed strong desire to participate in the implementation process. The level of participation of women is limited and negligible. However, the women members have constantly engaged themselves with the VWSCs and ensured that the process of implementation is not guided by vested interests of the socially and politically powerful.

A moderate scenario is observed in some other areas of the North Chotanagpur region. In one of the project villages, the dominant families of the village have captured all positions of office bearers that have been provisioned by the constitution. Exclusion is quite evident in these areas and therefore participation is limited. The SHGs, however, have been engaged with the VWSCs and through some representation in the committee they have influenced the VWSC's work. Subsequently, the women here participated in bookkeeping, accounting and supervision of works.

However, good practices in some pockets of Chotanagpur region were also observed. In Champadih (Koderma), the best case scenario in North Chotanagpur, the SHG women are more involved in group discussions within the VWSC. These women are therefore involved in project supervision and are more than willing to contribute to the process of various future activities such as collection of funds from users and supervision of works.

Thus, in the North Chotanagpur areas (non-tribal villages), women's participation are constantly challenged by the dominance of upper caste, influential families and the economically well off and male members of the village. It is also evident that these challenges can be overcome by external interventions that strive for equity, equal participation and social empowerment of the vulnerable. Participation of women is certainly related to the level of empowerment of the SHGs though their influence on the PRI system remains a challenge that would possibly require long-time engagement.

11.4.4 Perceptions about Water Security as Consumers

Women from all project sites have indicated that arranging water for drinking purposes of the household is exclusively their responsibility. This is irrespective of whether the population is tribal or non-tribal. They indicated that finding sufficient water to meet all their requirements is also very hard. On an average, a women (sometimes assisted by her adolescent girl child) have to go out of home to fetch water for six times in a day as households have a practice of using fresh water for their immediate requirements. Every time there is a need for water for cooking, drinking, bathing, or for homestead animals, women have to set out of home and cover an average distance of 1 to 2 km to fetch water. At home there are not enough storage facilities and even if there is storage facility the quantity of water that can be carried manually is limited to 10 litres per adult or adolescent. Thus, even if two women from a household are transporting water, they can bring home approximately 20 litres of water per trip. Thus, water required till midday is brought in the forenoon and water required in the evening or night is carried home in the afternoon.

Also, in some cases accessing safe sources is a challenge. In Rohne (Khunti) and in the villages of Dumka, women said that they access water from a stream in the forest area. They also reported that they have to be careful not to confront elephants in the forests. Women in Rohne have narrated agonising experience of hiding in the forests with small children while fetching water from a stream, when confronted by a herd of elephant.[4]

For the women, water security is a solution to the difficulties as narrated earlier. Regular supply of safe water in homes will save time and the hardship, and also eliminate the risks involved in fetching water. The FGDs revealed that the women accord immense value to the water supply programme; respite from the excruciating stress of fetching water being the primary reason. They dread the 'hardship' involved in fetching water, while they understand that there is no escape from that responsibility. As a social norm it is impossible for the men to replace the women in water duties.

Interestingly, the FGDs further revealed that the women would prefer to relax at home during the time that is saved, as water is made available at home. They do not have complex and gigantic aspiration to fulfil in the incidence of additional time available due to availability of water at home. Some women conveyed that they would take better care of their children or will be able to give more time to farming. In conclusion, it may be stated that the struggle for fetching water comes at a huge cost to the women as it affects their health, and child care at home and other productive involvements are compromised. The women, across the board, also acknowledged that running water at home would also secure the health of the household.

The men generally had a different perspective on water security. For the men in leadership, the initiation of the water supply scheme in the village was both a matter of pride and also professional achievement (as office bearer of the *Panchayat*). They indicated quite clearly during the FGDs that the advent of the water supply scheme actually enhances the profile of the village. However, they also appreciate the fact that the project will be fulfilling the long pending need of the women. The men could also identify the health benefits associated with safe water, that is, water available through piped water systems.

In both tribal and non-tribal populations, the male population agreed that the task of fetching water is designated to women. The male counterparts empathised with the difficulties faced by the women in arranging water for the household. In tribal populations, the male counterparts viewed this as a norm wherein they had little role to play, other than arranging for improved water supply systems, such as water source within the premise of the house.

In tribal populations, the men share very little responsibility in the household chores as a matter of their social norm. In an overall canvass of deprivation and underdevelopment, they stick to the traditional division of labour—which is limited to farming activities (or in the absence of farming work, any casual work they could perform). They agree to the risks involved in fetching water from forests or far-off places and opined that availability of water at home would be a boon to the household.

11.5 Summary and Conclusion

In this chapter, we studied community-based mini piped water supply schemes, in the context of women's leadership, participation and consumption of the drinking water supply schemes. We have analysed key areas that need addressing for future programming in the sector. The learning from this experience is used to develop the Standard Operating Procedure for implementation of the community-based mini-piped water supply schemes in Jharkhand. In summary, the following trends could be stated as critical characteristics in the women-water bond in rural areas:

- Leadership of women in the VWSCs is critical for addressing the aspiration of women in matters related to implementation and management of water supply schemes in rural areas. This representation is strengthened in the presence of SHGs in the same village. Optimum results are obtained if the SHGs overlap with the VWSCs.
- In cases where the SHGs and VWSCs are not aligned, the former still plays a role in the operation of the VWSCs. At the village level, the SHG women challenge the activities of the VWSC and ensure participatory processes wherein the women's voices are heard. In the long run, an agenda to align the VWSC and SHG would certainly ensure better results.
- In non-tribal population, male counterparts often represent women members nominated or elected in village-level institutions. The

representation of the family (through the male member) was an overriding priority. This is a barrier to manifestation of the women's aspiration in real terms.

- In tribal populations, women emerge as natural leaders. Again, women here do not contest for individual leadership—on the contrary they work collectively and provide leadership to their common aspirations. Therefore, access to benefits is equitable. There is no sense of privilege attached to individual leadership.
- In non-tribal areas, individual leadership is highly valued and contested; different layers of influencing factors such as caste, economic status, religion and gender, determine the leadership in these populations. Evidently, these populations are more vulnerable to exclusions, particularly the marginalisation of women.
- Women's participation in the tribal villages (Kolhan and Santhal Parganas) is not restricted by gender discrimination[5] as commonly witnessed in non-tribal populations; in tribal populations, women bear the brunt of their conventional role of being the central figure in the family coupled with continued deprivation and underdevelopment (Roy, 2012). In the villages of North Chotanagpur inhabited by caste Hindu or Muslim populations, gender discrimination is clearly manifested. In both cases, empowerment of women is a huge challenge; SHGs appear to be a fair response to the need of these women.
- The objectives of the SHGs are not limited to improvement in conditions of livelihood. It also addresses the general well-being of women and their families. In this context, 'water' emerges as a critical component for the women.
- In rural areas, fetching water for household needs is a women's responsibility as a matter of social norms. It is almost impossible for a man to replace the woman of the household for this responsibility. Men are sensitive to the risks and drudgery associated with fetching water but they can never really share this burden.
- Lack of water leads women to compromise with their health, well-being, child care and other productive ventures. Most critically it entails hard work, which drains them off their energy. Most women conveyed that if they can escape this, they would relax during the time saved. They would also contribute more effectively to other household chores and child care in case water is available at home.

These observations are critical for designing the institutional mechanism and defining the roles and responsibilities of key stakeholders who would be involved in project implementation. It emerges clearly that in a water supply scheme, the involvement of women is mandatory in order to ensure sustainability. Again in order to make the woman's participation effective, she needs to be adequately empowered against discrimination or vulnerability depending on the social fabric in which she exists. The SHG movement definitely is a key conduit for women's empowerment. Hence, the constitutionally provisioned VWSCs need to be supplemented by involvement of SHGs in the village. If SHGs do not exist, then attempt should be made to encourage SHG formation facilitated by competent agencies.

It is important to appreciate community dynamics and the processes through which women emerge as leaders or active participants in implementation of water supply projects. There is no magic solution to remove the barriers that prevent women to participate and lead rural development programmes. However, women's leadership and participation can be facilitated through a process of engagement and most critically the facilitator should be an expert in the field. Also, as indicated in the chapter, women's perception as user or consumer of drinking water needs to be understood realistically.

Hence, the key recommendations flowing from the aforementioned observations are primarily for the government departments/service providers. Given that the projects will be executed with government funds, the office bearers should be fully aware of the observations listed above and also take into cognizance the broad recommendations given below:

- The drinking water sector should proactively explore the possibility of involving SHGs in management of the schemes; wherever possible a strong functional relationship between the VWSCs and SHGs should be worked out.
- The drinking water sector should make conscious efforts to introduce 'drinking water (and sanitation)' as a key agenda for the SHGs to look beyond 'livelihood security' and strive for improvements in 'quality of life'. This requires advocacy work with agencies involved in SHG formation and women's participation.

- It is of paramount interest to explore and engage technical agencies involved in formation and advancement of SHGs and women's participation. The technical agency actually provides necessary inputs to ensure a functional linkage between VWSCs and SHGs and address field level issues related to their inter-relationships.

Acknowledgements

The authors acknowledge Sarbani Bose, Abhijit Mullick, Niranjan Kumar, Raju Maity, Devmalya, Navin, Ramesh, PRADAN, Manisha Mehra, Rohit Sharan, Jitendra Agarwal, UNICEF, Jharkhand, and Mahesh Agarwal, Action for Community Empowerment (ACE).

Disclaimer

The views expressed herein are those of the authors and do not necessarily reflect the views of UNICEF or the United Nations.

Notes

1. Tank/pond/lake, river/canal, Spring.
2. Third scheduled tribes have been notified in Jharkhand by the State Government.
3. Less than 20 per cent of the population in the North Chotanagpur area is tribal as against 20–80 per cent in other regions of the state (Census, 2011).
4. Tribal communities traditionally have close contact with wild animals, being dependent on forests for their livelihood. It is said that 'tribal and tigers' have lived in harmony for years; however, in the present-day context, these are being perceived as security issues in the tribal communities.
5. Adivasi Women: Situation and Struggle—Seminar paper presented by Sangarshrath Adivasi Mahila Manch, published in People's March, Vol. 7, No. 7, August–October 2006.

References

Census. 2011. Government of India. Available at http://www.census2011.co.in/ (accessed on 10 October 2014).

Mitra, Jayanta. 2011. Piped water flowing in each house in Belkhera: A journey towards social cohesion, Community Based Water Supply in Jharkhand—UNICEF & DWSD, Government of Jharkhand.

Prakash, Amit. 1999. Contested discourses: Politics of ethnic identity and autonomy in the Jharkhand region of India. *Alternatives; Social Transformation and Humane Governance*, 24(4): 461–496.

———. 2007. Case Study 'Tribal Rights in Jharkhand', UNDP supported by Asia Pacific Gender Mainstreaming Programme (AGMP).

Roy, Debjani. 2012. Socio-economic status of Scheduled Tribes in Jharkhand. *Indian Journal of Spatial Science*, 3(2, winter issue): 26–32.

Sangarshrath Adivasi Mahila Manch. 2006. Adivasi women: Situation and Struggle. *People's March*, 7(7).

Sharma, Subhas. 2004. Tribal development in Jharkhand—A multidimensional critical perspective. *Studies of Tribes and Tribal*, 2(2): 77–80.

SECTION 3

Case Studies: Sanitation

12

12.1 Introduction

12

Enabling Gendered Environment for Watershed Management

Eshwer Kale and Dipak Zade

12.1 Introduction

Women play a significant role in natural resource management, particularly in water resource. Watershed development is one of the major programmes in India, which focuses on ecosystem regeneration and water conservation. Women's participation is widely recognised in watershed programmes; however, it is not addressed beyond activities like SHG formation and conventional credit and savings. Even though women are involved in *Shramdan*[1] and physical work, they are not given equal opportunities in decision-making, particularly in addressing concerns related to domestic water, fodder and fuel.

Watershed Organisation Trust (WOTR) with its long-term engagement with rural communities evolved a new approach for watershed development called the 'WASUNDHARA approach'. *Samyukta Mahila Samiti* (SMS)[2] is an integral component of watershed development programme in WASUNDHARA approach. In few of the WOTR's watershed projects, Integrated Domestic Water Management (IDWM) project was implemented, which addressed domestic water and sanitation (WATSAN) concerns in watersheds. This chapter discusses these two interventions and their gender implications.

Five detailed case studies of SMS from different parts of Maharashtra were selected and another research study was carried out to evaluate the IDWM project. The results highlighted that these interventions provided institutional space and opportunity for women to build capacities and confidence to take lead in addressing issues of domestic water, health and sanitation. For this, financial autonomy for women is necessary in programmes. When these interventions are intertwined with community-managed and resource-strengthened programmes like watershed development, their effectiveness and sustainability is further enhanced. Further, these models succeed in addressing equity and inclusion concerns effectively.

Hence, we recommend that with few relevant modifications, SMS as an institutional tool and IDWM project model, which effectively addressed water, health and sanitation concerns of the community through women's active participation should be integrated in the current Integrated Watershed Management Programme (IWMP). This will effectively address women's concerns.

12.1.1 Water Access, Availability and Changing Roles of Women

Water gives life, and therefore it is the basic requirement for the ecosystem's sustenance. India has 16 per cent of the world's population and 4 per cent of fresh water resource (UNICEF, FAO and SaciWATERs, 2013). Per capita annual freshwater availability declined in India from 5,177 m^3 in 1951 to 1,545 m^3 in 2011. This decline is a major concern because a country is considered water stressed when per capita availability of water drops below 1,700 m^3/person/year (Khurana and Sen, 2010). Today, we are much below this scarcity figure. At the national level, nearly 60 per cent of all districts in India have problems related to either the quantitative availability or to the quality of groundwater or both (Cullet, 2012). Along with the increasing water scarcity, the domestic water accessibility and affordability to the marginalised and poor sections of society are also major concerns. In the present rural context, domestic water issues are intrinsically related to women. In India, women are one of the most vulnerable and overburdened sections of the society. The issues of gender bias and inequity point to the double subjugation of poverty and gender inequity (Prakash, 2003). Although, women's role in natural resource management and water-related activities is clearly visible, they are mostly excluded from resource ownership and in the decision-making process of developing and managing these resources and services (Wahaj, 2007). However, as women are central to family, agriculture and village life, and have the potential to influence the change processes, the need for women's participation is widely recognised in water management and WATSAN programmes. This chapter focuses on dealing with some of these issues outlined earlier through an innovative approach developed by WOTR in the western Indian state of Maharashtra.

This chapter is structured as follows: In the initial introductory section, we discuss the scope and importance of watershed development in India and the process of emergence of women participation in it. Then we give the details of WOTR's interventions, *Samyukt Mahila Samiti* (SMS) and Integrated Domestic Water Management Programme (IDWM) and how their specific gender perspectives developed. We share the detailed findings and data of the selected interventions in relations to impacts of the SMS and IDWM and show how the project helped to change their lives. In the concluding section, we summarise the major findings and recommend that there is urgent need to incorporate the SMS and IDWM interventions in mainstream watershed programmes in India.

12.1.2 Scope and History of Watershed Development in India

Approximately 65 per cent of all agricultural land in India is rain-fed (Joshi et al., 2005). It was anticipated that watershed-based eco-restoration programmes could effectively meet the emerging and complex challenges, namely high poverty, unemployment and acute degradation of natural resources. Watershed development is one of the largest single development initiatives undertaken in terms of resources, geographical spread and agencies (governmental and nongovernmental) involved (Lobo, 2005; Planning Commission of India, 2012).

Traditionally, watershed programmes in the country were supply-driven. Often, such approaches did not match the needs of stakeholders in the watershed in the absence of their participation (Joshi et al., 2005). Over time, the nature and scope of watershed programme has undergone considerable modification and a more participatory approach to watershed development has been adopted. The first generation watershed programmes (1970–1980) focused on soil conservation and catchment protection of reservoirs, while second generation watersheds (1980–1994) focused more on water conservation and improvement of irrigation and moisture conservation (Joshi et al., 2005). Successful watershed projects of the early 1980s such as *Ralegan Siddhi, Adagaon* and *Pimplalgaon Wagha* in the state of Maharashtra and Sukhmajori in Haryana, as well as Participatory Integrated Development of Watershed (PIDOW) projects in Karnataka laid a foundation for participatory approach in watershed development.

Community participation became part of the mainstream programmes of the central and state governments with the emergence of the 1994 guidelines issued by the Ministry of Rural Development. Further revised Hariyali guidelines (Government of India, 2003) by the Ministry of Rural Development and Ministry of Agriculture made effort to institutionalise participation. With increasing need of watershed development, several non-government organisations (NGOs) actively participated in implementing this programme and demonstrated the importance of people's involvement in the success of watershed development. The third generation watershed projects (post-1994), however, revolved around the participatory approach by involving the stakeholders in the planning, implementing, monitoring and sharing the benefits and costs. The recent Common Guidelines for Watershed Management, 2008, issued by the Government of India, also give central focus to community participation (Government of India, 2008).

12.1.3 Emergence of Women's Participation in Watershed Development

Women's participation is an extremely important concern in watershed development programmes; however, for a long time it did not enter the discourse. Many watershed project guidelines and various committee reports talk about the equity concerns in the watershed. The Eswaran committee (Government of India, 1997), which looked into the training and capacity building in the watershed development after the 1994 common guidelines, expressed concerns for equity as one of the goals of watershed development. The 1994 common guidelines also stated that it was necessary to give special emphasis on improving the economic and social conditions of the poor and the disadvantaged sections of the watershed community such as those without assets and women (Government of India, 1994).

Until now, most of the provisions and activities discussed in context of woman's participation are either about reserving seats for them in village-level watershed committees (VWC) or village development committees (VDC) and mandatory formation of women SHGs. However, field observations and findings of numerous studies show that these provisions proved either insufficient or inadequate to make women visible and to involve them effectively in decision-making processes in watershed projects.

12.1.4 Women's Role in Drinking Water, Sanitation and Health

Government of India has made huge efforts towards providing safe WATSAN facilities through various central and state initiated programmes starting with National Rural Drinking Water Supply programme in 1969. The second generation programmes started with the Technology Mission in 1986, which was later renamed as Rajiv Gandhi National Drinking Water Mission in 1991–1992. The third-generation programmes started with the sector reform process in 1999–2000, where the approach for provisioning services shifted from being a 'supply driven' to 'demand driven'. Community participation and contribution was the key component here; these programmes were the Total Sanitation Campaign in 1999 and Swajaldhara in 2002. Currently, the fourth phase has begun with major emphasis on sustainability of water, equity in access and decentralised approach with the involvement of *Panchayat Raj* Institutions and community organisations (Plan International, 2009; Government of India, 2010).

Despite these efforts and multitude of programmes, the present WATSAN situation in rural India is not very encouraging. About 74 per cent of the rural habitations/villages have been declared as fully covered (with 40 litres per capita day water) by drinking water facilities, whereas, sanitation coverage in rural areas is only 30.7 per cent (Plan International, 2009; Government of India, 2012). Even households that are fully covered suffer from issues like water quality, slippage from fully covered to partially covered, problem of source sustainability and rapidly declining water table due to excessive groundwater extraction. Universal access to and usage of sanitation facilities coupled with poor hygiene practices are other key challenges.

Inadequate or lack of WATSAN facilities not only has health-related implications but also economic consequences. A study by Pattanayak et al. (2010) states that inadequate WATSAN services, the second most common cause underlying medical conditions that lead to child mortality, impose considerable illness, and coping costs on households in developing countries. These costs fall disproportionately on the poor, and impact women and children. According to WATSAN Programme of World Bank, in India the total economic impact of inadequate sanitation amounts to ₹2.44 trillion a year. This is equivalent to 6.4 per cent of India's GDP in 2006 (World Bank, 2010).

Given the above background, this chapter documents the approaches by WOTR, which has been exercised for the last two decades as integrated watershed development programme in rural semi-arid regions of Maharashtra. These approaches—SMS as an institutional tool for empowering women and IDWM—effectively addressed water, health and sanitation concerns of community through women's active participation and have made considerable changes in the lives of the people.

12.2 WOTR-WASUNDHARA: New Approach for Developing Watershed

WOTR is working since 1993 on watershed development. It has implemented 1,265 projects in six states (either direct involvement or provided capacity building/financial support). WOTR has been following WASUNDHARA approach for implementing watershed development programmes since 2005. 'WASUNDHARA' is a Sanskrit word, which translates as 'the Earth' and connotes compassion, caring, co-responsibility and harmony. It takes around five years to complete the watershed project. Each project team consists of a technical expert/officer, woman social worker and social development officer, along with watershed community.

This approach is based on the following four principles and components:

1. Village envisioning entails a Village Action Plan (VAP) developed by the village to map out the trajectory of their development: where

things stand, what are the objectives of individuals and the village as a whole and how might these objectives be reached.

2. Addressing equity is an important part of this concept. Each social and economic group in village is proportionally represented in the VDC and families contribute to development activities in proportion to their economic standing, with the poorer families paying less than the better off.
3. Gender inclusiveness focuses on overcoming traditional gender barriers by promoting women's involvement in the village development process. The three essential elements of gender inclusiveness are (a) 50 per cent of the VDC seats are reserved for women on a rotational basis, (b) women's SHGs and SMS are formed in the village, and (c) women in the VDC, SHGs and SMS are trained on the tools needed for them to be active decision-makers in the community, including training to manage funds and accounts for women's development activities.
4. The sustainability of the project is achieved through capacity and institutional building. The project is implemented through the *Gram Panchayat* (GP) by the officially recognised VDC. The VDC and other community-based organisations (CBO) are trained in project management, monitoring and evaluation.

As on date, this approach has been adopted in more than 200 projects implemented by WOTR and its partner NGOs.

12.2.1 Samyukt Mahila Samiti

Watershed development has been the focus area of the WOTR since its inception. In the early stages of WOTR's work in Indo-German Watershed Development Programme (IGWDP), it was soon realised that women were not only benefiting less from watershed development, but were disproportionately overburdened and distressed. Understanding the problem, women were organised through SHGs, along with other village-based institutions such as VWC. These SHGs were federated in an apex body called as SMS. Though savings and microfinance were the major components of these SHGs, it was aimed to build confidence in women to be active stakeholder groups of the project. Therefore, SMS was the major component of the earlier IGWDP guideline implemented by WOTR and is an important component of the present WASUNDHARA approach. At present, SMS is one of the major strategies of gender inclusiveness in the watershed projects.

SMS: structure, formation, activities and provisions: From the beginning of the project, women are encouraged to participate in the planning and decision-making in the activities. Around the fourth month of the project, the WOTR team starts helping women to form SHGs, which generally consist of 15–20 women who come together to do savings and organise credit activities. Around the seventh month of the project, each of these SHGs nominates one or two members to form part of the SMS. Thus, SMS is the apex body of women who represent the various SHGs formed within the watershed, covering various hamlets. Then, they proceed to manage development activities directly linked to women, such as drudgery reduction activities, health and sanitation (toilets, awareness programmes, etc.). The SMS members are also active in village-wide discussions (including *Gram Sabhas*) on the development process as a whole.

During the initial capacity-building phase and full implementation phase (FIP), there is a provision to allocate 5 per cent of the total sanctioned project budget amount for women's developmental activities with a view to reduce drudgery of women. This women's development fund is released to SMS's bank account. SMS also receives training immediately following its formation and during various project phases. It also receives one day training towards the end of the project to emphasise on the continuity and sustainability of the SMS and explaining how to register the SMS as a legal body. The women development fund from the project to SMS bank accounts is the incentive for long-term continuation of SMS meetings and activities.

12.2.2 IDWM Project

The IDWM component was implemented by WOTR in nine villages from Aurangabad, Dhule, Pune, and Ahmednagar districts of Maharashtra. All these villages are predominantly tribal. The total number of households covered through this project was 631 and the total population was 4,194.

The project, supported by Arghyam Trust (Bangalore), was implemented during 2008–2011. WASUNDHARA approach was followed during the implementation of this project. The key objectives of IDWM project were to improve the drinking WATSAN condition in project village and to create an IDWM model for further replication. Along with these the project was aimed to promote safe disposal of domestic wastewater and to promote awareness among the village communities for judicious use of water through water budgeting tool. The selected villages were from drought prone areas having severe domestic water problem. Drinking water was supplied through tankers during summer months.

The IDWM project was implemented in various phases varying from need assessment to project evaluation. The first phase focussed on planning, capacity building and community mobilisation. The Domestic Water Supply and Sanitation Committee (DWSSC) was the main actor of the IDWM project. It was responsible for planning and implementation of WATSAN activities, maintaining assets created under the project and developing VAP. The project funds were transferred directly to the DWSSC's bank account. DWSSC is considered as a sub-committee of GP as per Bombay GP Act 1958 under section 49(1). It works under the control and supervision of GP and consists of equal number of women and men (at least one-third GP members as per the act).

The members of DWSSC, including the members of other sub-committees such as Social Audit Committee (SAC), purchase committee and supervision committee were selected in *Gram Sabha*. The second phase included implementation of the project wherein private toilets were constructed in all the project villages. In many villages, it was difficult for BPL families to pay for their contribution. In such cases, WOTR facilitated the process by linking these families to the government subsidy schemes.

In all villages, government drinking water scheme already existed. However, it was suffering from many problems such as leakage, inequitable distribution and maintenance. WOTR's main interventions in this area focussed on digging and construction of well/borewell, laying pipelines connecting households, construction/repairing of overhead water storage tank, installation of electric pump and electric connection, source strengthening, and providing household tap connections or public stand posts. Families were convinced and motivated to either make a kitchen garden or to construct a soak pit. The water budgeting concept was introduced in the villagers with the view to sensitise the village community towards judicious use of water and most importantly for ensuring domestic water availability during summer season. A simple water quality testing kit was provided in all villages. Additionally, drinking water samples from these villages were tested in government's water testing laboratory. These test reports were routinely discussed in the SHG and SMS meetings. As a result, women started purifying drinking water at home.

The third phase of this project included monitoring and evaluation of the programme using qualitative assessment matrix (QAM) and peer group assessment (PGA). QAM is a tool wherein the community assesses its own achievements and lacunae facilitated by the implementing organisation. For PGA, people from one village evaluate the quality of work in other village. The objective of doing this assessment was to impart monitoring and assessment skills to community and facilitate learning from other's experiences. The assessment team had equal representation of women.

12.3 Assessment of SMS and IDWM Projects from a Gender Perspective

In order to understand these two approaches and their impact on the lives of the women, a study was initiated to evaluate the gender-inclusive strategies of the WOTR's interventions in watershed projects. Further, the study tried to understand the impacts of IDWM project on WATSAN condition and women's role in making the projects effective.

12.3.1 Methodology

Five detailed case studies of SMS were selected from different regions of Maharashtra. These case studies were selected keeping in mind the effectiveness of SMS and its evolution during different phases and project guidelines. The second intervention this chapter discusses is the IDWM project. A small qualitative research study was conducted in 2011 to evaluate its effectiveness. For this purpose, five villages were selected from nine IDWM project villages. These villages were: Chawadipada and Chafaban (Dhule district), Wanjarwadi (Aurangabad district), Satichiwadi

(Ahmednagar district) and Thapewadi (Pune district). In each of these five villages, detailed interviews from each of the following four respondent groups were taken:

- The President/Secretary of DWSSC
- A key person from a local institution (health committee/SMS/VDC/GP/primary school teacher)
- An adult male villager from village (who is not a member of any local institutions)
- An adult female villager from village (who is not a member of any local institutions)

In addition to the above mentioned 20 interviews, detailed interviews of the project team leader (TL) and the social development officer were conducted to better understand the organisation's perspectives on gender issues. Thus, a total of 22 interviews were conducted.

12.3.2 Domestic Water: Changes in Access and Availability

Prior to the IDWM project, most of the villagers were not getting sufficient domestic water throughout the year and women were the prime sufferers. The difficulties faced by women regarding water scarcity prior to the project are clubbed in four main categories: travelling long distances and spending much of the morning time on fetching water, physical hardship, social problems (like conflicts) and lack of personal hygiene. Of these, travelling long distances/spending more time was reported as the main difficulty resulting in physical pain for fetching domestic water. This is well illustrated by one of the women respondents from Thapewadi village:

> I had severe abdominal pain on the day of delivery of my child. Still I fetched water from 6 am to 10 am and made ten trips. I delivered the baby at 2 pm on that day.

The water scarcity has led to prevalence of poor personal hygiene conditions. One of the women respondents reported,

> Having a daily bath and washing clothes were a problem earlier. We used to take bath and wash cloths once in 2–3 days. The teachers used

to scold children for not having regular bath. (Woman respondent from Chawadipada village)

Villagers also reported social problems such as inequitable domestic water access. In water scarcity period, the poor and women were forced to fetch water from private water sources (dug wells), which many times resulted in facing humiliation and conflicts. As reported by a respondent,

> Earlier we had to fetch water from a far-away well. The owner of the well quarrelled with us and threw away our pots in well. He verbally abused us. (Woman respondent from Wanjarwadi village)

After the project implementation, water availability has increased and regular water supply is ensured for all. The attendance of children at schools is increased and they are neatly dressed. In general, improvement in personal hygiene, reduction in time and distance for fetching water, and decline in social problems and physical hardship are some of the changes mentioned by the respondents.

Followed by reduced time/distance, improvement in personal hygiene is the major advantage mentioned by people. The president of DWSSC from Chafaban village shared how the project contributed in their life:

> Lot of time has been saved for women due to the water scheme. Cleanliness of people has improved. They now live a decent life.

Women shared that now they spend this time on other domestic chores such as taking care of children and spending leisure time with neighbours and friends. The case study from Wankute watershed village in Box 12.1 illustrates how women and SMS took initiatives in addressing their drinking water and other problems in the village.

12.3.3 Improvement in Sanitation and Hygiene Condition

Prior to IDWM project in the villages, open defecation was the general practice. Very few households had toilet facilities and even these were not in use regularly due to water scarcity. Travelling long distances and

Box 12.1:
Wankute village in Sangamner taluka

Wankute village is situated in Sangamner taluka of Ahmednagar district. One of the most important social outcomes of the project has been the involvement of women and the contribution they have made to make their lives simpler and better. Through participation of 142 women in nine SHGs, SMS took initiative in project to reduce drudgery and improve the quality of lives. Women were trained on issues concerning their health, growth monitoring of their children, nutrition using local resources, personality development, etc. However, beyond just training, the women took interest in availing other necessary services to make their life easier. With the ban on felling trees for fuel, with WOTR's assistance women purchased solar lamps, hot water *chulhas* and smokeless *chulhas* to reduce their dependence on wood as a source of fuel. Wankute watershed has very undulating topography (highly hilly) and settlement is scattered. One of the hamlets known as Nandale Pathar consisting of around 25 tribal households is situated at the upstream (on a sloppy hill). It is around 1.5 km from the main village settlement (*gaothan*). For domestic water requirement, women and children from these households were forced to come down and again climb up the hill, carrying water pots on their heads. This was time-consuming, risky and extremely painful for the women and children. While working in the village, WOTR realised the problem and with active participation of the community (SMS and VDC) constructed a dug well for drinking water in the Nandale Pathar at the upstream of watershed and resolved water issues.

inconvenience resulted in difficulties for old and sick people, pregnant women and children. Also, going out in the dark with the fear of insect bites and snakebites was common. Further, a respondent from Satichiwadi village stated: 'Women's dignity was at stake because of open defecation.' It was difficult for pregnant women to travel long distances.

Another major concern shared was the lack of privacy for defecation reported by almost all respondents. Few respondents reported health risks because of houseflies, mosquitoes and the dirty water percolating down into the wells situated downstream. Some respondents also mentioned about incidences of skin rashes as a result of not maintaining personal hygiene.

Post-IDWM project, people reported that the villagers are experiencing the advantages of toilet construction such as convenience of using toilet at any time, sense of privacy, reduction in time spent and distance travelled for defecation, and perceived improvement in health and cleaner surroundings. As stated by SMS president of Wanjarwadi village below, the sense of privacy while using toilets was a major advantage.

'We are free from the shameful situation because of toilet construction. We are happy and satisfied about it.' Respondents reported that incidences of ailments like dysentery and malaria have reduced by half compared to earlier situation. Now, there are lesser visits to doctors, and money is also saved. The case study from Kachner Tanda No. 2 in Box 12.2 illustrates how the SMS took initiatives in addressing their drinking water problems and constructing bathrooms for each household.

Box 12.2:
Kachner Tanda No. 2 in Taluka and District Aurangabad

Kachner Tanda No. 2 is one of the hamlets of Kachner village in Aurangabad district of Marathwada region. The watershed project in the hamlet was implemented during July 2008 to March 2012. The SMS, named as *Hirakani* has 14 members, which represents seven SHGs from the village. The SMS initiated *Gram Swachhata Abhiyan* (total sanitation campaign) as an entry point activity in their village. Training was provided to SMS on record maintenance and management of the group. They have participated in exposure visit to villages Naralewadi, Hivrebajar, Ghulewadi and Mhaswandi where they learned how SHGs work, and they witnessed business initiatives like dairy farm, grocery shop and biogas for cooking purposes. With WOTR's guidance, SMS was formed, which mobilised women for watershed project-related issues like ban on free grazing, tree cutting and ban on liquor sale in the village. The SMS became very active in the village when they played an important role in the implementation of various activities such as construction of bathrooms and solar home light system. As generally in villages people do not invest in private bathrooms, women face the problem of lack of privacy for bath and personal hygiene. Therefore with the support of WOTR, the SMS took lead and constructed separate private bathrooms for each household (54) in the village. Because all families were facing severe drinking water problems and they had to fetch water from distant private or public wells, SMS laid a drinking water pipeline with tap connections to each household. The group members are proud that all the developmental efforts were done collectively, and they have realised their potential in village development.

12.3.4 Impact on Livelihoods and Quality of Living

The IDWM project has improved their quality of life, villagers conveyed. According to them, the way of living has changed. In their words, now people get sufficient water to take bath regularly and wear clean cloths. Villages are now cleaner due to safe disposal of waste water. Earlier, garbage would be thrown anywhere and nobody cared for it. Now, it is disposed at one place.

> Life has become happy and better. Women are free from the worries related to fetching water. Earlier they could not even sleep properly due to this worry. Now they are free from the worry. (Woman respondent from Wanjarwadi village)

The case study in Box 12.3 depicts how SMS helped women to boost confidence by promoting livelihood opportunities and changed their quality of life.

Box 12.3:
Kachner Tanda No. 6 in Taluka and District Aurangabad

Suman Lahu Jadhav is a resident of Kachner tanda no. 6. She is 30 years old and illiterate. Her family consists of her son, two daughters and her husband. The family stays in a *kaccha* house. Their source of income for this family was mainly agriculture and some meagre earnings through daily wages.

In 2008, when WOTR started women's empowerment activities as a part of watershed development intervention, it changed Suman's life. She became a part of the 13-member SHG, 'Damini'. She started contributing ₹50 monthly towards the group saving. Three years earlier, Suman took a loan of ₹5,000 from the SHG at 3 per cent annual interest rate, and started a grocery shop in the village. Initially, the shop started with selling small items like sweets, soap sachets, snacks, coconut oil and such items. Within two years, she bought a refrigerator to preserve and sell cold drinks in summer. She also started selling other necessary groceries like sugar, food grains, soaps and biscuits. Currently, Suman earns a weekly net profit of around ₹4,000. Suman and her husband make all the purchases for the shop. She now has plans for further expansion of the shop.

12.3.5 Capacity Building and Institutional Representation of Women in CBOs

Box 12.4:
Bhoyare Pathar, in Parner Taluka, Ahmednagar

In the following case study, the initiative taken by the SMS in Bhoyare Khurd village in Ahmednagar district provided the inspiration to the women in other villages. The IGWDP in Bhoyare village began in 1996 and ended after five years of implementation in 2001. The project has two main phases: Capacity Building Phase (CBP) and FIP. In CBP, to access the community's participation for FIP, a micro watershed (around 250 hectares) of the larger watershed was taken for implementation on an experimental basis. For the first time in the history of the programme, SMS took the lead in taking the overall responsibility to implement the micro watershed, and, so, only women through SMS successfully implemented it. This initiative not only helped women in the village for building confidence in them, but also proved the benchmark to women in surrounding villages to come forward and take lead in watershed development. Along with the implementation, SMS helped many families through microfinance to overcome the economic constraints to undertake micro-businesses (such as goat keeping and poultries) and to meet family expenses related to health, education and marriage.

The DWSSC president of Chawadipada village stated that 'If women had not taken the lead, then I do not think that the project would have been completed.'

A male respondent from Wanjarwadi village metaphorically explained women's role in the project as 'We sat in the bogies and made women sit in the engine. Women were the drivers and men were the assistants.'

The project succeeded in mobilising women for village development. As a result of women's involvement in the project activities, there have been significant changes in them. As reported by a male respondent from Chawadipada village, 'Earlier women would not come together for any type of work. But now even if one SHG member is informed about a meeting, she gathers other women quickly.'

12.3.6 Ensuring the Sustainability of the Project Impact

Sustainability of project activities and impact for longer period is the key challenge in development sector, the key behind the success of IDWM project and SMS in addressing women's concern and WATSAN issues is that these interventions are intertwined with overall watershed project. IDWM project was implemented in only those villages where integrated watershed projects were implemented by WOTR. This watershed project implementation, with its increased resource availability and social capital (institutional building), provided the basic platform on which the IDWM and WATSAN activities are based. Therefore, most of the respondents reported that watershed development has resulted in increased water availability and groundwater table. They have sufficient water even during summer months. One of the respondents stated that VDC, SHGs, SMS and other local-level institutions formed as part of watershed project helped in collective decision-making abilities and enhanced unity. These attributes helped in creating an ideal platform for implementation of IDWM project. According to the project TL, as a result of watershed development, water is recharged and its sustained availability is ensured. Further, the concept of water budgeting would not have much acceptance in the community without watershed development. This is because the efforts of the whole community is involved in making a watershed successful and hence judicious and sustainable use of available water by whole community is ensured. SMS, which evolved in IGWDP and further revised in WOTR's WASUNDHARA approach, is the major institutional tool for increasing gender inclusiveness in watershed projects.

12.4 Conclusion and Recommendations

The IDWM project experience has shown that women can effectively participate in WATSAN programmes and make them successful. In order to accomplish this, they need training in leadership and organisation. Source protection and sustainability are key issues in any WATSAN programme. The SMS case study experiences show that if appropriate and adequate opportunities and institutional spaces are created for women with capacity-building strategy and financial autonomy, they can effectively address their concern, especially the problem of domestic water availability and accessibility, health and sanitation. The findings show that when these interventions (IDWM and SMS) are intertwined with community-managed and resource-strengthened programmes like watershed development, their effectiveness and sustainability is enhanced. Further, these models have addressed issues of equity and inclusiveness as it covered all households and all sections of the community, including the very poor and the vulnerable groups getting due representation and benefits in the programme.

Our experiences show that, at institutional level, persuading men to accept women's leadership is quite challenging, therefore trainings on gender sensitivity with follow-up meetings (for men and women) may prove effective in this direction. Women members from poor households struggle to attend meetings and trainings as it means they will lose their daily wage. Women who have never been active in their village are also often quite afraid to take on responsibilities. Hence, confidence building and support from the project team is necessary. Having explained the relevance of the SMS and IDWM project, we argue that for meaningful inclusion and participation of women in watershed development, separate spaces such as sub-institutions or sub-committees (such as SMS, village health committee) with financial autonomy and well-defined decision-making process are necessary. WOTR's experiences show that this approach strengthens the bargaining position of women within the watershed community, as compared to their nominal participation in VDC and SHGs. IWMP, which is the umbrella programme under which the major watershed programmes are brought together in 2008, lacked these provisions.

We recommend that with few relevant and appropriate modifications, SMS as an institutional tool and IDWM project model, which effectively addressed water, health and sanitation concerns of community through women's active participation should be integrated in IWMP to effectively address women's concerns in watershed development and management.

Acknowledgements

We thank Thomas Palghadmal from WOTR for his inputs throughout the study and Dr Hemant Apte for providing research guidance for the

IDWM study. We also thank Anuradha Phadtare, Ashish Pardhi and Joseph Shinde from WOTR for their assistance during the fieldwork and sharing their experiences.

Notes

1. Shramadan is the voluntary offering of labour for development activities.
2. Watershed village level federation of all women self-help groups.

References

Cullet, P., Paranjape, S., Thakkar, H., Vani, M.S., Joy, K.J. and Ramesh, M.K. 2012. Groundwater—towards a new legal and institutional framework. Forum for Policy Dialogue on Water Conflicts in India, 2012, pp. 58–71.

Government of India. 1994. Watershed guideline—1994. Available at http://www.watergovernanceindia.org/acts.php?cat=national&cat_id=11 (accessed on 30 September 2012).

———. 1997. Report of the committee on training for watershed development (Eswaran committee). Report. Available at http://www.indiawaterportal.org/sites/indiawaterportal.org/files/Eswaran%20Committee_Training%20for%20Watershed%20Development_MoRAE_1997.pdf (accessed on 6 March 2013).

———. 2003. Guidelines for Hariyali-2003. Available at http://dolr.nic.in/hariyaliguidelines.htm (accessed on 30 September 2012).

———. 2008. Common guidelines for watershed development projects-2008. Available at http://dolr.nic.in/CommonGuidelines2008.pdf (accessed on 20 September 2012).

———. 2010. National rural drinking water programme: Framework for implementation. Department of Drinking Water Supply Report Department of Drinking Water Supply Ministry of Rural Development. Available at http://rural.nic.in/sites/downloads/pura/National%20Rural%20Drinking%20Water%20Programme.pdf (accessed on 10 November 2012).

Government of India, Ministry of Home Affairs. 2012. Availability and type of latrine facility: 2001–2011. Available at http://www.censusindia.gov.in/2011census/hlo/Data_sheet/India/Latrine.pdf (accessed on 1 April 2013).

Joshi, P.K., Jha, A.K., Wani, S.P., Joshi, L. and Shiyani, R.L. 2005. Meta-analysis to assess impact of watershed program and peoples participation report. ICRISAT, Andhra Pradesh, India. Available at http://www.lk.iwmi.org/brightspots/PDF/South_Asia/MetaAnalysis.pdf (accessed on 2 August 2012).

Khurana, I. and Sen, R. 2010. Drinking water quality in rural India: Issues and approaches. Report Water Aid. Available at http://www.wateraid.org/documents/plugin_documents/drinking_water.pdf (accessed on 20 August 2012).

Lobo, C. 2005. Reducing rent seeking and dissipative payments: Introducing accountability mechanisms in watershed development programs in India, Paper presented in World Water Week, Stockholm (August 2005). Available at http://www.wotr.org/upload/2.Reducing%20Rent%20Seeking%20and%20Dissipative%20Payments.pdf (accessed on 1 October 2012).

Pattanayak, S.K., Poulos, C., Yang, J.C. and Patil, S. 2010. How valuable are environmental health interventions? Evaluation of water and sanitation programs in India. Report. Available at http://www.who.int/bulletin/volumes/88/7/09-066050/en/index.html (accessed on 15 November 2011).

Plan International. 2009. Evaluation of existing capacities in WATSAN sector. Report. Plan International, India, p. 10. Available at http://www.ddws.gov.in/sites/upload_files/ddws/files/pdf/pdf/EvaluationExisting.pdf (accessed on 12 September 2012).

Planning Commission of India. 2012. Minor irrigation and watershed management for the twelfth five year plan (2012–2017). Report. Available at http://planningcommission.nic.in/aboutus/committee/wrkgrp12/wr/wg_migra.pdf (accessed on 15 October 2012).

Prakash, D. 2003. Rural women and food security. Rural Development and Management Centre, New Delhi, India. Available at http://www.uwcc.wisc.edu/info/intl/rur_women.pdf (accessed on 10 January 2012).

UNICEF, FAO and SaciWATERs. 2013. Water in India: Situation and prospects, 105 pages. Report. Available at http://indiasanitationportal.org/sites/default/files/WaterinindiaUNICEF%20Report.pdf (accessed on 15 April 2013).

Wahaj, R. 2007. Gender and water. Securing water for improved rural livelihoods: The multiple uses systems approach, IFAD, Rome. Available at http://www.ifad.org/gender/thematic/water/gender_water.pdf (accessed on 1 September 2012).

World Bank. 2010. The economic impacts of inadequate sanitation in India. Report. Water and Sanitation Programme, World Bank. Available at http://www.wsp.org/wsp/sites/wsp.org/files/publications/wsp-esi-india.pdf (accessed on 15 November 2012).

13

Women-led Total Sanitation: Saving Lives and Dignity

Gregor von Medeazza, Megha Jain, Ajit Tiwari, Janardan Prasad Shukla and Nisheeth Kumar

13.1 Need for Improved Sanitation for Women and Children

13.1.1 Background

Open defecation has profound effects on women's life in India, where 65 per cent of the population still does not have access to improved sanitation (WHO and UNICEF, 2013). There are significant adverse impacts of open defecation on women's dignity and privacy. These include holding back urination and defecation until darkness; dangers of exposure to weather conditions, insects and snakes; dangers of sexual harassment and shame of exposure. Additionally are risks of potential health complications.

Furthermore, evidence reveals that globally, 88 per cent of diarrhoeal deaths are directly attributable to unsafe drinking water, inadequate sanitation and poor hygiene conditions (WHO, 2004). Open defecation is one of the major sources of microbiological contamination of water sources and drinking water leading to diarrhoeal diseases. In India, diarrhoea is among the largest killer diseases and accounts for 12.6 per cent of child deaths in the country (Liu et al., 2012). Furthermore, there are links between open defecation and malnutrition (Chambers and von Medeazza, 2013; Spears, 2012a and b), which is a leading cause of child mortality and morbidity in India and an impediment to the cognitive development of children (Dillingham and Richard, 2004).

13.1.2 TSC's Shortcomings

Government of India, with its Total Sanitation Campaign (TSC) launched in 1999 has been struggling to address the issue of sanitation with tools like information, education and communication (IEC), financial incentives and capacity building. Although 275 million people in India have gained access to improved sanitation between 1990 and 2011, around 620 million still defecate in the open (WHO and UNICEF, 2013). India's proportion of open defecation in the world has risen from 51 per cent in 1990 to 55 per cent in 2000 and around 60 per cent in 2011 (WHO and UNICEF, 2013). The wealthiest 40 per cent of Indians are 10 times more likely than the poorest 40 per cent to use improved sanitation (Narayanan et al., 2011). After more than a decade of programme implementation and fund disbursement, Census 2011 reported the national sanitation coverage at 31 per cent (Government of India, 2012). This represented a gap of 57 million 'missing' toilets and raised serious concerns about the approach taken by the campaign, its implementation plan and monitoring system built under the TSC. TSC's poor performance was officially acknowledged in late 2011 when the then Minister of Drinking Water and Sanitation stated that the 'Total Sanitation Campaign has been a failure. It is neither total, nor sanitation nor a campaign' (Tandon, 2011).

Similarly, in the state of Madhya Pradesh, while the TSC reported over 70 per cent sanitation coverage in rural areas, the Census 2011 found only 12.2 per cent of households with improved sanitation. Evidence from the field in Madhya Pradesh revealed that either toilets built under the TSC were only 'reported' to be built or were partially built with faulty technology. It was also reported that toilets were built and put to other uses such as store rooms, etc.

13.1.3 Need for Collective Behaviour Change

Though the TSC was designed to be community-led, people-centred, incentive-based and demand-driven leading to total sanitation, it eventually turned out to be government-led, hardware-centred, subsidy-based and supply-led instead. As an alternative, sanitation practitioners have successfully experimented with the Community Led Total Sanitation (CLTS) approach to accelerate the elimination of open defecation. CLTS[1] focuses on collective behaviour change for toilet *use* rather than solely *construction* of toilets and to sustain open defecation free (ODF) status of communities. It uses a set of carefully and intelligently adaptive motivational tools and techniques to facilitate the community's own analysis of their defecation practices, and trigger collective action to become ODF, without depending on hardware subsidies from government and external agencies.

13.1.4 Objectives of the Chapter

The key objective of this chapter is to enable learning and provide replicable practices on how to encourage the emergence of women natural leaders and steer their role in catalysing ODF outcomes and community development at large through the cascading effects, which emerge in the CLTS process. The selected case studies from ODF villages of Guna and Budhni, in the Indian state of Madhya Pradesh presented hereafter will support this objective by showing how women-centred sanitation movements provided a conducive environment for accelerating the elimination of open defecation at community level. The final section of the chapter will draw inferences and summarise the key lessons learnt for scaling-up and replication.

13.2 CLTS in Madhya Pradesh: Experiences and Lessons Learnt

With UNICEF's support, several block administrations in the state of Madhya Pradesh (most notably Guna and Budhni) have set themselves the target to make their blocks ODF. They adopted the CLTS approach, which offer interesting insights and learning for women empowerment.

CLTS approaches were started in 2009 and 2011 in Guna and Budhni blocks, respectively. CLTS triggered community consciousness to end open defecation and stirred a movement based on pride and dignity and need for better hygiene. In most cases the key hook, which triggered communities, was the issue of *izzat* (respect) for women who exposed themselves with shame every day while defecating in the open along with the risks of sexual harassment. CLTS triggered exercise that combined the *individualistic approach* and the *community approach*. Successful ODF villages chose the second path where women played a crucial role in convincing, motivating and constructing toilets. This was primarily because they are the worst affected and most marginalised to articulate their discomfort. In the process of creating an ODF community, the output and outcome has not been limited to '100% toilet construction' and 'declaring the village ODF,' respectively. The impact of the CLTS approach on women stakeholders has been profound in some villages. The success stories are now spreading fast to other villages who are keen to become ODF. This compounding effect in terms of achieving more and more ODF villages is bringing about a silent social change for better quality of life where women motivators are in the lead.

By the end of August 2013, more than half of the villages in Budhni block achieved ODF status through UNICEF-supported CLTS programme (Figure 13.1).

Similarly, around 30 villages have become ODF through UNICEF-supported CLTS in Guna block. CLTS implementation in Guna and

Figure 13.1:
Number of villages in Budhni block, MP, achieving ODF status up to August 2013

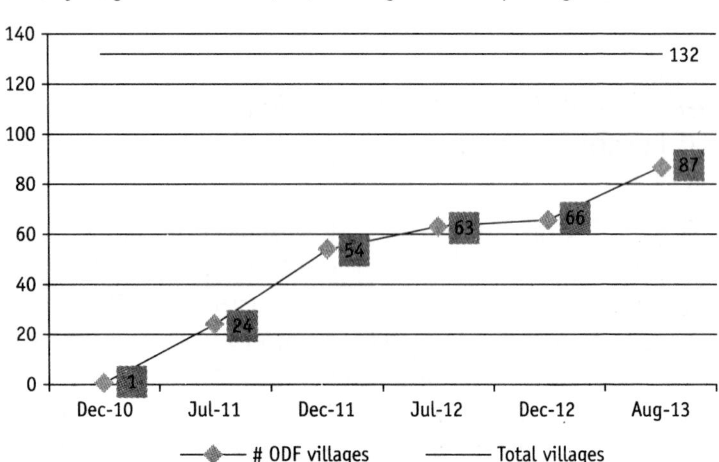

Budhni has helped women liberate from age-old social and gender constructs and enhance their self-respect and dignity to lead an empowered life. The role of women in bringing social change through CLTS has been duly recognised at the policy level and stirred government action in the form of the new state-level sanitation campaign, MARYADA (translating as 'dignity' from Hindi to English), launched in 2012. The state of Madhya Pradesh indeed adapted the national *Nirmal Bharat Abhiyan*[2] (NBA) to be a women-centred sanitation campaign, using the CLTS approach as the backbone of its implementation. This indicates a visible shift in government thinking towards the adoption of community-based approaches to address engrained problems of water and sanitation as against traditional subsidy-based development schemes.

13.3 Experience and Lessons Learnt from Guna

Thirty-five villages of Guna became ODF following the CLTS approach between May 2011 and now. These are remote villages of Guna block in district Guna, where communities, particularly women, took the lead in making their villages ODF by collectively helping people in realising the ill-effects of poor sanitation. They mobilised people for the collective purchase of construction items, and gained basic masonry skills using local material. They also helped others in constructing their toilets and ensured the sustainability of behaviour change through morning vigilance teams to check on defaulters. They now continue their work within their communities beyond the achievement of the ODF status. The communities in remote villages in Guna, mostly inhabited by scheduled tribes, have been the area of operation of these women natural leaders who emerged from the CLTS programme. Gradually they started moving towards villages inhabited by general castes. Another unique feature of the Guna example is that this CLTS programme led to the creation of an NGO (formed and registered under the Indian Societies Act in December 2012). This was entirely formed by the natural leaders, including women, who emerged during the CLTS process. Despite being poor, they have demonstrated tremendous ability and commitment and making several villages ODF. The CLTS movement in Guna is taking the shape of sustainable and scalable programme.

The achievement of 35 ODF villages in Guna district with focus on collective behaviour change in the given time span is significant, particularly in view of the fact that not a single village had become ODF in the district despite 8 years of TSC implementation between 2003 and 2011. It is noteworthy that TSC focused on subsidies and construction of toilets and not on agents of change, particularly women. The following case study illustrates this.

13.3.1 Devgarh Village of Sirsi Gram Panchayat: *Lab of Innovations*

13.3.1.1 Village Background

Devgarh has 123 households inhabited over three hamlets of Devgarh, Phooti Bawadi and Mehra spread over 5 kilometres. The Pateliya tribe inhabits the village. There were no functional toilets. Since the village is on the way and close to Nihal Devi temple, which attracts many pilgrims it makes the village more vulnerable to open defecation by outsiders. Diseases such as diarrhoea and malaria were rampant and women had to face inconvenient situations due to the practice of open defecation. Many water sources, particularly hand pumps were not properly functioning; those that did had stagnant pools of water surrounding them with broken platforms.

13.3.1.2 Addressing the Problem of Open Defecation

Triggering was undertaken on 12 February 2012 in Devgarh hamlet; 12 natural leaders emerged, out of which three were women. Post-triggering, two monitoring committees were formed, one comprising women headed by Wardi Bai and the other of men headed by Himmat Singh. When the village was triggered, the harvesting season was at its peak and households had dearth of money to spare for toilet construction. However, the triggering stirred a sense of shame and disgust, and the need to eliminate open defecation was deeply internalised. In only two days, 38 pits were dug for toilet construction through self-labour. By the sixth day, 10 households had arranged for materials like toilet pans, p-traps and pipes, and another 40 households had pooled ₹20,000 (US$400) for collectively buying the materials. The landlord willingly extended loans; labour contributions were made by natural leaders for reluctant families; sand for the mortar was taken from the river bed and local materials were used for the superstructures.

Four women natural leaders emerged and played an active role in CLTS implementation. As members of the *Nigrani Samiti* (Sanitation Monitoring Committee), they undertook early morning monitoring rounds at 4 a.m. during the winter months with a whistle and torch, to check out for open defecators. Construction of toilets became complete in every single household on 17 March 2012. The committee members decided to initiate child vigilance on open defecation. Total 150 children were divided into teams and different areas were allotted to them for early morning vigilance. The teams were provided with whistles, which they would blow once they catch anyone defecating in the open. The practice was continued for a few more days. Soon no one flouted the rule and the village became officially ODF on 21 March 2012, barely 37 days after the initial triggering had taken place. The villagers continued their vigilance to catch outsiders during the holy Navratri celebrations. Natural leaders of Devgarh put up banners at different places along the roadside proclaiming their village as ODF and warning outsiders not to defecate in their area. They also constructed a couple of additional communal toilets for outsiders and provided them water and soap for hand-washing and toilet flushing.

13.3.1.3 Result and Impact

Devgarh's natural leaders formed teams to help other villages in stopping open defecation. One of the woman natural leaders, Wardi Bai, proudly says:

> Despite having been an elected Sarpanch (village secretary) for 5 years, I was not able to talk to anyone; but now I can easily speak-up and have the courage to address anyone, not only in my own village but also in other villages. I don't feel shy any longer; I am proud now.

Similarly Parvati Bai says: 'Until I had not joined this movement, I was unhappy and alone. But after joining, I have forgotten the pain of being a widow.'

Parvati Bai now confidently goes around with a Kanni (plastering blade) in her hand and politely asks a person not taking interest in toilet construction: 'Tell me what is restraining you. How can I help?' Surprised by the widow's confidence to help, community members of other villages still defecating in the open get ashamed at their helplessness and their made-up excuses. Parvati Bai explains how harvesting and toilet construction can be done simultaneously by sparing a few hours and how not much money is required. She willingly extends help for masonry work as well. She goes around neighbouring villages narrating the story of her village where people helped each other and are now proud of their achievement.

Lalita Bai, a widow with four children, was estranged by her family and locked up in her house. Since the time she joined the sanitation campaign, her self-respect has been boosted. She has now reunited with her family members who take pride in her and treat her with respect.

These women leaders have played an important role. Post-ODF, 90 households in Devgarh village built compost pits and now dispose cow dung and other bio-degradable waste in these pits. With their acquired skills in masonry, the women also extended help in digging compost pits and promoting solid and liquid waste management. Also, water being typically a women's issue, hand pump maintenance was also undertaken as an after-effect of the CLTS movement when hygiene was inculcated as a culture. Women natural leaders also contributed in the repairing of hand pumps, filling boulders, creating cement platforms, etc.

Furthermore, Devgarh's school operates now more regularly and with higher quality, and enrolments rates have increased. Self-help groups (SHGs) have been formed for enhancing household savings and a club for women farmers was formed to ensure that at least one acre of land per household is used for organic farming. The larger vision of these empowered women is to 'make our villages, Panchayat, cluster, block and entire district Open Defecation Free. Until the time all villages are not ODF, we and our children are not safe: the root cause of all problems is shit.' It is to achieve this vision, that the natural leaders of the area created a formal NGO in December 2012 to help rendering their entire Guna district ODF.

13.3.1.4 Sustainability and Replicability

After rendering their village ODF, the natural leaders have actively contributed in rendering another 11 neighbouring villages ODF. These were achieved through their experience sharing during triggering sessions, follow-ups, early morning visits, supporting the organising of women's group, conducting meetings, taking out rallies for awareness and providing direct help in toilet construction. These women natural leaders have helped directly in the construction of toilets in 20 villages and are capable of constructing a full toilet (sub- and super-structure) independently, without any support from men. By now, out of the 286 villages of Guna block, this group of women natural leaders from Devgarh has gone to some 80 villages for experience sharing and CLTS triggering activities.

13.3.1.5 Lesson Learnt

The case study of Devgarh village offers useful insights for implementing CLTS in large and scattered villages. Women's participation has influenced the district administration to recognise and support people's collective efforts. While Sumer village of Puraposar *Gram Panchayat* was the first village to become ODF in Guna block, Devgarh village of Sirsi *Gram Panchayat* was a major breakthrough as it led the movement in the entire block by contributing to a number of innovations that were followed by other villages.

13.4 Experience and Lessons Learnt from Budhni

Until 2009, the case of Budhni was similar to Guna, where the traditional construction-based approach to implement the TSC by doling subsidies to BPL families was being followed. IEC activities were being undertaken as a compulsory budget line activity without any actual effect on behaviour change for creating open defecation communities. CLTS was introduced in Budhni in 2010 with the support of UNICEF. A team of 50 CLTS facilitators (divided in 5 teams of 10 people each) was trained and constituted, and a block-level action plan was prepared to meet the set target of creating 138 ODF villages in 62 *Gram Panchayats*. However, about 30 facilitators gradually lost interest because of the extreme field effort involved and left the initiative. With steady progression, 14 CLTS facilitators remained (divided in three teams) and worked in 30 selected *Gram Panchayats* on the west side of the block. Under the leadership and firm resolve of the CEO, block administration Mr Ajit Tiwari, the CLTS efforts were continuing with limited manpower and progressive success. One unique aspect of the CLTS initiative in Budhni has been the fact that it is entirely government-led with only marginal support from UNICEF and CLTS agencies. This is also one of the reasons why the initiative has faced various contextual problems, which have been tactfully mitigated through local solutions and leadership techniques, including political will. Support for the CLTS initiative was gathered at the onset by inviting high-level officials in the first CLTS training of trainers workshop. It also included motivation for subsidy at all levels of governance within Budhni block. Here, subsidy was instituted as a performance-based incentive, which was given to communities only after rendering their respective community ODF. It was clearly articulated by the block administration that no subsidy would be given for toilet construction before open defecation was eliminated. Furthermore, communities' mistrust on government officials and PRI functionaries was mitigated by making regular and repeated visits by block-level functionaries during early morning hours to monitor open defecation. *Gram Panchayat* functionaries were strategically kept at distance during CLTS implementation to avoid charges of corruption and money embezzlement with regard to release of subsidy. At the same time, *Asha* and *Anganwadi* workers were engaged effectively to target women for collective behaviour change. The block CEO and his team conducted a series of tailored training sessions with these village-level functionaries to orient them on CLTS. The purpose of these trainings was to clarify the linkage existing between sanitation, hygiene, health and nutrition for women and children. A veiled *Anganwadi* worker who leads the women's Sanitation Monitoring Committee team in Mogra village confidently replied when asked how she finds time to lead the CLTS initiative in her village, while sanitation is not part of her main job responsibility: 'I have to do it. Sanitation cannot be separated from my core function; it's my village and my people.' Strong interpersonal communication methods were used to trigger the community into action and refuse the practice of open defecation.

13.4.1 Case Study from Ratanpura Village: A Scalable Approach

13.4.1.1 Village Background

The small village of Ratanpura at the foothills of Vindhyachals in Budhni block of Sehore district in Madhya Pradesh had one toilet out of 113 households until December 2010. The village was unclean with an overhanging pungent smell of open excreta and flies all around. Diarrhoea, fever and other related ailments were common amongst villagers, and a large proportion of the households' income was spent on medical treatments. Women were exposed themselves at the loss of their self-respect and dignity. The time of menstruation was further worsened with the pain of travelling long distances and difficulty in maintaining proper hygiene.

13.4.1.2 Addressing the Problem of Open Defecation

Against this backdrop, the block-level functionaries, who were trained in the CLTS approach, entered Ratanpura village with the aim to render it ODF. These functionaries were well aware of the plight of women and approached them to discuss the issue of ending open defection in their village. During a village-level meeting, women gradually opened up and started sharing their experiences:

Sukhmaniya Bai, an old woman, said, 'In the rainy season, open defecation is so challenging and risky; I have to walk so long, and often my clothes get dirty.'

Rama, 16 years old, said, 'It is really disgusting to go near the main road and defecate there; we have to stand-up every time someone passes by and then sit down again. Most of the time we are unable to clean ourselves properly; sometimes, young boys laugh at us.'

The women's own responses and experiences of open defection acted as a trigger to instigate them in taking action to change their present situation of shame, humiliation, pain and disrespect, and probe for a solution. The women reached a consensus that they, in fact, are the worst sufferers and they no longer want to continue with the same practice of having to defecate openly.

After an initial discussion with women, the CLTS functionaries/facilitators also included the men in the triggering session. During these sessions, Geeta Bai emerged as a natural leader. She said, 'When we bring a daughter-in-law in our house, then we use "parda" (veil covering the face) to save her dignity, but the very next day we are sending her outside the house to defecate, for everyone to see her shitting half naked in the open. Tell me, where is then our "parda" and dignity!!' Many other women followed Geeta Bai and narrated their everyday woes and fears of defecating in the open.

Once the villagers were faced with questions of women's issues and shame, the CLTS facilitators suggested that they could teach the villagers to construct simple pit latrines as an immediate solution. Next day, five women started digging pits at dawn and completed the construction of five latrines by noon. This was the start of a movement in the village where toilet construction was now proven to be a daylong job, which could also be done by women! The women leaders who emerged during the course of this process formed monitoring committees and established a village action plan to end open defecation in their village, with the help of the CLTS facilitators. Several households started building simple pit latrines, and in the span of 10 days, all households of Ratanpura village had completed their pit latrine.

However, the men were still defecating in the open. All this time, the CLTS facilitators were also raising awareness on the sanitation and health linkages, which proved to be the next motivational trigger for women to convince their husbands also to stop open urination and defecation. The monitoring committees started supervising common defecation areas, at the early morning hours with whistles to catch defaulters. Radha Bai hilariously narrates, 'The "Nigrani Samiti" (monitoring committee) members started whistling when my husband went for open defecation. So, eventually, my husband decided to build a toilet and use it.'

13.4.1.3 Result and Impact

On 26 January 2011, the village became 100 per cent ODF. At a village meeting, some villagers decided to improve their simple latrines to low-cost leach pit toilets. The CLTS facilitators helped them understand the leach pit design and guided the construction activities with suggestions in making it cost-effectively so that every household would be able to afford it. Women leaders came forward to learn some basic masonry skills, which until then were considered men's work domain. Many offered their services free of cost. The village headman and other functionaries, and other local PRI members also supported the activities. Group procurement was encouraged to avail bargains with local suppliers; interest-free loans were facilitated; local sand was used to reduce material cost; material donations were also forthcoming. Chhote Lal from Ratanpur village said, 'The Sarpanch arranged for sand and cement bags. They also arranged for loans from the brick kiln.'

Ratanpura villagers now proudly boast of their self-achievement in making their village ODF. The villagers recognise the overall improvements since they have eliminated open defecation in their village, and the benefits they have accrued in terms of health, cleanliness, social cohesion and their increased respect towards women. The women of the village narrate, 'There have been no flies and mosquitoes this year. Fever, cholera, diarrhoea was worse in our village earlier.'

Men convey, 'We feel good that women speak-up now. They didn't go out the house much. Now they come here (in meetings).' At the community level too, there has been a significant change in thinking: 'Now we sit together as one community and talk about important communal issues such as the improvement of the road, the drain, mending the fences for cattle, etc. We come together and decide how to improve these faster.'

13.4.1.4 Sustainability and Replicability

As the success story of Ratanpura spread fast to the neighbouring villages, and many visited the newly turned ODF village to see the changes. Women from Ratanpura left the confinements of their homes and personally went to share their experiences with other villages. The experience which women from Ratanpura gained during the CLTS implementation process also helped in shedding many myths on issues such as the fact that: 'toilets are expensive'; 'it is the government's duty to provide toilets'; 'one can fall in the toilet pit and suffocate or die.' The women of Ratanpura are now active change agents in bringing further development in their villages, beyond ODF, and building a better future for themselves, their children, their family and their community as a whole.

13.4.1.5 Lesson Learnt

The case of Ratanpura offers insights on how the administration can assess needs and turn them into development opportunities using women-centric strategies. The effects of CLTS have not just been limited to ending open defecation and achieving health, hygiene and comfort for women. It has also affected women's life in many subtler ways. There is a reported significant reduction in disease burden; everyday familial and social stress, community cohesion and ability to work in teams has increased; class and caste divides have narrowed; skills in planning and management, public relations, public speaking, negotiation and communication, have been enhanced; there is an increased connect between community and administration for better governance with emphasis on women-led and community-based projects.

Apart from the above-stated impact-based learning, there was also a number of process-based learning that may be replicated. Ratanpura showed how the fight against open defecation could lead women to express their needs and claim for their rights, providing them basic skills to facilitate solution finding and decision-making, providing them with avenues to execute their ideas. Women emerge as effective leaders when such an environment is created. The case study also illustrates that women-centric projects help in nurturing women leaders who can then take on greater responsibilities in the future, entailing positive effects on the overall socio-economic development of their communities. Women natural leaders are often able to enrol women functionaries in their team building and implementation exercise thus having a rippling effect of accelerating women's empowerment.

13.5 The Way Forward: Recommendations for Scaling-up CLTS in Madhya Pradesh

This chapter presented women-led sanitation movement in the state of Madhya Pradesh, which led to the elimination of open defecation in over 100 communities across the state. The case studies presented in this chapter showed how women's leadership has a vital role to play at all levels in the campaign for *Nirmal Bharat*, a clean and ODF India. The main lesson learnt of these case studies is that whenever and wherever possible, women should be identified, triggered and supported to lead the drive for ODF and hygienic conditions, including menstrual hygiene.

The field learning and emerging experiences from UNICEF-supported CLTS initiatives in Guna and Budhni provided a credible platform for the Government of Madhya Pradesh to now replicate and scale-up its successes. The CLTS field demonstration in those two blocks served as the impetus for the Government of Madhya Pradesh to build its state-level sanitation policy around the principles of women-centric and community-led. The state sanitation policy in the form of the MARYADA campaign ('women's dignity') now serves as the national NBA's application in the state of Madhya Pradesh (Government of Madhya Pradesh, 2012).

MARYADA has been tailored to give emphasis on creating a women-centred sanitation movement, with woman dignity at its core, through community-led approaches, with CLTS as the 'operating system'.

The MARYADA guidelines also draw from the recently approved National Sanitation and Hygiene Advocacy and Communication Strategy (SHACS), with regard to the four key hygiene behaviours (UNICEF and MDWS, 2012), underline the 'beneficiary's involvement in construction' and leave sufficient room for flexibility of design. The guidelines also include school and *Anganwadi* centre sanitation, emphasising on child's right as well as O&M. MARYADA's first phase focuses primarily on the state's 5,800 piped-water supply scheme villages, deriving from the strong correlation existing between sanitation and household water availability.

However, key challenges remain, such as the discrepancy between reported data of toilets constructed on the one hand, and actual number of households with toilets used. Progress in toilet use has remained slow in the state, despite exponentially increasing funds. To achieve ODF communities through CLTS, extensive and well-trained human resources are needed, especially, women natural leaders who as we have shown, play an important role to accelerate the elimination of open defecation.

To overcome these challenges, the following recommendations are to be considered. Firstly, greater emphasis is needed on demand generation for toilet *use*. Secondly, the sequencing of funds released to communities is key: pre-ODF subsidies have the risk to hinder community-led approaches, undermining self-help, focusing monitoring on toilets constructed rather than ODF achievement; allocations should thus to be given to communities only after communities have reached ODF status (as post-incentive, not pre-subsidy). Thirdly, the monitoring mechanism of toilets *constructed* versus toilets *used* and ODF status must be strengthened.

Finally, as we have argued in this chapter, collective behaviour change can be accelerated through facilitating ownership of the communities, and a desire for dignity and self-respect. The elimination of open defection can come through self-help in a few weeks, as seen in the case studies presented in this chapter, even in lagging states such as Madhya Pradesh, through the implementation of community-led approaches and provided that their key principles and sufficient capacitated human resources are in place on the ground. In order to achieve its vision of a *Nirmal Bharat* (clean, ODF India) by 2021, the NBA must constitute itself an 'army of sanitation foot soldiers,' exclusively commitment to the improvement of sanitation across the country. Ideally, one trained sanitation frontline worker should be identified in each hamlet to carry out interpersonal communication at household and community levels for the elimination of open defecation through collective behaviour change. At least half of these sanitation frontline workers should be women; the representation of women must also be increased at all levels of the bureaucracy and within the Government's decision-making system in charge of sanitation.

This chapter has shown that the impact of poor sanitation on women is far greater than they are for men. The opportunities for engaging more with women and women's movements are therefore huge. The national sanitation policy should thus redesign its strategy to articulate and incorporate a well-defined and constructive role for women, and for engaging with existing women's movements, SHGs, forums and platforms, for contributing, awareness raising, advocacy and implementation of women-centric collective behaviour change.

As a concluding remark, this chapter has shown how over the last couple of years, CLTS in the state of Madhya Pradesh has not only proven to be a potentially successful approach to render communities ODF in a sustainable manner, but also how women have and can increasingly emerge as empowered natural leaders. Women's roles in communities go way beyond intended sanitation objectives. (Refer to Annexure 13.1, depicting the conceptual 'Women's empowerment ladder through CLTS'.)

Disclaimer

The views expressed herein are those of the authors and do not necessarily reflect the views of UNICEF or the United Nations.

Notes

1. CLTS was developed in late 1999 by Dr Kamal Kar in Bangladesh. He has taken the leading role in spreading CLTS to many other countries (for more readings on CLTS refer to: www.communityledtotalsanitation.org; Kar and Chambers, 2008). CLTS has now been implemented in over 50 countries,

some 20 of which have embraced CLTS without subsidy as national policy. As a result, an estimated 25 million people are now living in over 40,000 ODF communities worldwide. In India, Himachal Pradesh stands out as an encouraging example by achieving the highest rural sanitation coverage jump in the decade, from 28 per cent to 67 per cent (Government of India 2012) through campaigns, CLTS and delaying the TSC subsidy to become a community reward once an entire habitation/village became ODF.

2. The national *Nirmal Bharat Abhiyan* (NBA) is the successor of the TSC, which emphasizes on community engagement and behaviour change for achieving sanitation outcomes. For more details: http://tsc.gov.in/TSC/NBA/AboutNBA.aspx

References

Chambers, Robert and Gregor von Medeazza. 2013. Sanitation and stunting in India: Undernutrition's blind spot. *Economic and Political Weekly*, 48(2): 15–18.

Dillingham, Rebecca and Guerrant L. Richard. 2004. Childhood stunting: Measuring and stemming the staggering costs of inadequate water and sanitation. *The Lancet*, 363(9403): 94–95.

Government of India. 2012. Census of India 2011. Houses, Household Amenities and Assets. Latrine Facility, Government of India.

Government of Madhya Pradesh. 2012. MAYRADA Guidelines, State Water and Sanitation Mission, Department of Panchayat and Rural Development, Government of Madhya Pradesh.

Kar, Kamal and Robert Chambers. 2008. *Handbook on Community-Led Total Sanitation*. Institute of Development Studies, Sussex, and Plan UK.

Liu, Li, Hope L. Johnson, Simon Cousens, Jamie Perin, Susana Scott, Joy E. Lawn, Igor Rudan, Harry Campbell, Richard Cibulskis, Mengying Li, Colin Mathers and Robert E. Black. 2012. Global, regional, and national causes of child mortality: An updated systematic analysis for 2010 with time trends since 2000. *The Lancet*, 379: 2151–2161.

Narayanan, R., H. Van Norden, A. Patkar and L. Gosling. 2011. Equity and Inclusion in Sanitation and Hygiene in South Asia. A Regional Synthesis Paper (UNICEF, WSSCC and WaterAid), Kathmandu.

Tandon, A. 2011. Sanitation campaign has come a cropper: Jairam. Retrieved 19 December 2012 from http://www.tribuneindia.com/2011/20111022/nation.htm#6

Spears, Dean. 2012a. How Much International Variation in Child Height Can Sanitation Explain? Rice Working Paper, 10 December 2012.

———. 2012b. Effects of Rural Sanitation on Infant Mortality and Childhood Stunting: Evidence from India's Total Sanitation Campaign, Rice Working Paper, 3 July 2012.

UNICEF and MDWS. 2012. Sanitation and Hygiene Advocacy and Communication Strategy Framework 2012–2017, UNICEF and Ministry of Drinking Water and Sanitation, Government of India.

WHO. 2004. Hutton, Guy and Laurance Haller, Evaluation of the costs and Benefits of Water and Sanitation Improvements at the global level, World Health Organization, Geneva.

WHO and UNICEF. 2013. Progress on Sanitation and Drinking Water: 2013 Update, Geneva and New York WHO/UNICEF Joint Monitoring Programme for Water Supply and Sanitation.

Annexure 13.1:
Women's empowerment ladder, through CLTS

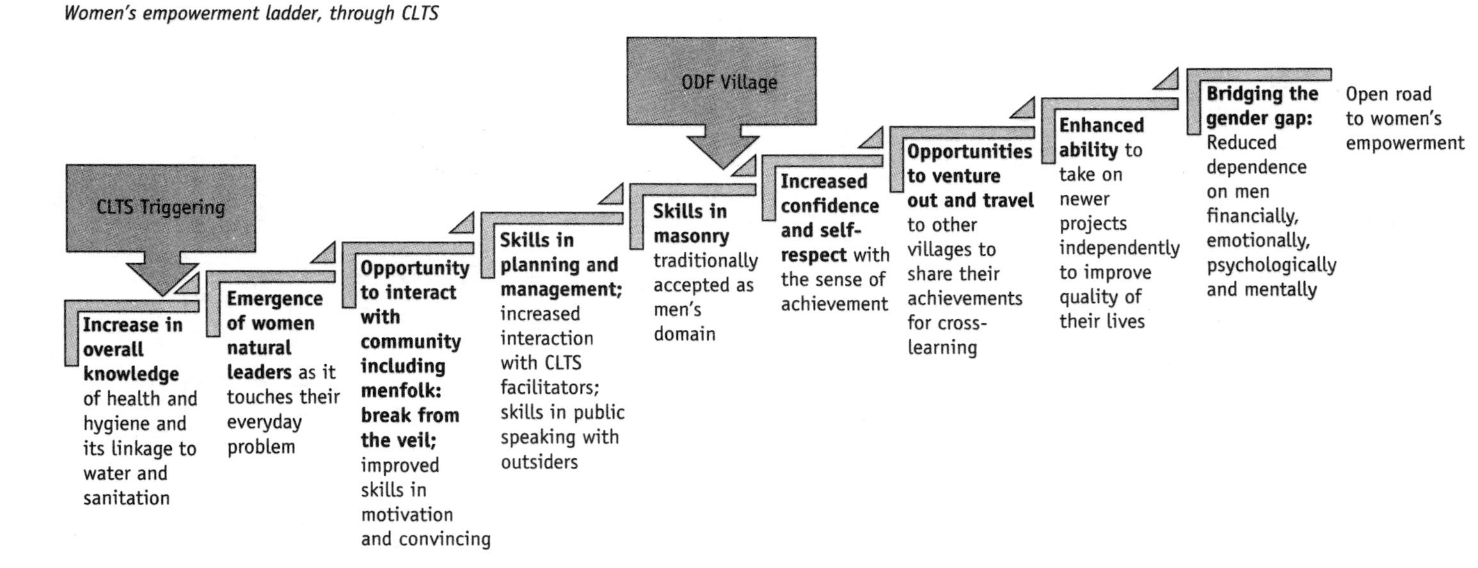

14

Innovative Approaches in Communication

Ajit K. Saxena, Shailesh Mujumdar and Gregor von Medeazza

14.1 Introduction

Safe drinking water, sanitation, and hygiene (WASH), fundamental to human development, is still not available to a large part of the population in India (WHO and UNICEF, 2013). An estimated 88 per cent of diarrhoea cases, the second most important cause of under-five child mortality, are related to poor WASH services (Hutton and Haller, 2004). Though India's Millennium Development Goals' (MDGs) target for access to drinking water has been officially achieved, India's MDG target for sanitation is seriously off-track. About 53 per cent of the Indian population still defecates in the open (Government of India, 2012), accounting for around 615 million people not having access to a toilet (WHO and UNICEF, 2013). Despite the fact that 259 million people in India have gained access to improved sanitation between 1990 and 2010, India's proportion of open defecation in the world has continued to rise, reaching around 60 per cent in 2011 (WHO and UNICEF, 2013). An estimated 212,000 Indian children under the age of five die each year from diarrhoea (Liu et al., 2012); strong correlations exist between open defecation and malnutrition (Chambers and von Medeazza, 2013) and the national economic loss due to lack of improved sanitation was equivalent to 6.4 per cent of India's GDP in 2006, accounting for US$53.8 billion (WSP, 2011).

Lack of adequate sanitation also causes indignity as women have no private place to attend to their menstrual hygiene management needs (EPW, 2012). Typically, women and girls place higher value in the need for WASH facilities than men. Yet the prevailing approaches for improving WASH services are not designed to address the strategic needs of women, and they are often excluded in the planning, management and monitoring process for WASH provisions. The exclusion of women is due to the fact that most traditional sanitation-related trainings and planning is carried out off-site, outside of the village and/or in a classroom-type mode. They are not adapted to the skills, capacity, interest and the work schedule of women. As a result, behaviour change and communication (BCC) efforts fail to get desired levels of community's attention and participation, and typically exclude women.[1]

UNICEF's snapshot on equity and gender in the WASH sector in India (UNICEF, 2012) underlines the need for strengthening the gender dimension across the sector. Furthermore, the sectorial review carried out jointly by UNICEF, FAO and SaciWATERs on the situation and prospects of WASH in India, discussed the role of each major stakeholder in overcoming current problems of the sector (UNICEF, FAO and SaciWATERs, 2013). In this comprehensive publication, special attention was given to the role of women, stating that the WASH sector still fails 'to recognise the fact that women can contribute immensely to the decision-making process by their vast unique experience, and therefore they should not only be seen as potential beneficiaries but also as actors' (UNICEF, FAO and SaciWATERs, 2013: 55). In the same vein, this chapter argues that conventional WASH approaches often fail to address the strategic needs of women and the key role they can, and indeed ought to, play in the planning, management and monitoring processes. We present a new approach and argue that prevailing approaches of awareness generation and capacity building for improving WASH facilities often lack forward and backward linkages in the supply and service chain. The challenges of innovation for awareness creation, capacity building and support systems to transform these stark figures of WASH in the light of varying topographical, climate, culture and economic backgrounds of the communities need to be addressed. Moreover, most approaches focus on single or limited aspects such as triggering, supply or marketing, hand washing, toilet construction, etc. and often fail to offer a holistic approach for WASH services covering all key components such as menstrual hygiene, construction, supply/services and governance/institutions in a comprehensive manner.

This chapter presents a new and innovative approach 'Pan in the Van' for capacity building and promotion of inclusive WASH services.[2] The 'Pan in the Van' approach has been tailored for providing women-centric WASH-related capacity building. The approach has been developed by

Energy Environment and Development Society (EEDS) with support from UNICEF. It is based on 'on-site' principles, keeping women at the core of the entire process, which is suitable to their context. It empowers women to take a leadership role. The 'Pan in the Van' approach rests on the premise that capacity building is the process whereby a community equips itself to undertake the necessary functions of governance and service provision in a sustainable manner, covering aspects related to skills, knowledge and attitude. It offers new ways on how women and children are enabled to become integral to sanitation improvement through playful awareness and exercises. Besides a set of original BCC games that actively engage participants, the 'Pan in the Van' also contains technical tools specifically focused on the capacity building of basic masonry skills for low-cost toilet construction.

The chapter draws from the main findings and lessons learnt from the 'Pan in the Van' approach in 80 *Gram Panchayats* (smallest unit of local governance in India) in five districts of Madhya Pradesh, between 2009 and 2012. During that period, 80 three-day long 'Pan in the Van' camps were carried out. An analysis of these interventions shows the improvement in message retention, behavioural change, involvement of women in planning their village WASH-related components as well as increased children participation in school sanitation.

The objective of this chapter is to present an innovative women-centric approach for the improvement of WASH services. The chapter will first outline the 'Pan in the Van' approach, its strategy and implementation process. It will then discuss some quantitative and qualitative findings of its field testing. The final section concludes the lessons learnt.

14.2 The 'Pan in the Van' Approach and Strategy

To address the issues and shortcomings outlined in the introduction of the present chapter, the NGO 'Energy Environment and Development Society' introduced, in collaboration with UNICEF, the 'Pan in the Van'[1] approach. The aim of the approach is to accelerate the elimination of open defecation through the Total Sanitation Campaign (now revamped as *Nirmal Bharat Abhiyan*, NBA) by addressing the gaps prevailing in capacity building and behaviour change communication efforts and is women-centric. It creates space for women's participation and empowers them to take leadership roles to become active in local governance and act as change agents. The 'Pan in the Van' approach was shortlisted as one of the 'promising approaches' by the World Water Forum during its 6th Conference held in Marseille, France (March 2012).[2]

'Pan' denotes the hardware and supply chain components, mounting into a 'Van', equipped with an audio video aid, a package of IEC tools and games, a plethora of technology options, exhibits, as well as a resource team for behaviour change (see Table 14.1 for a detailed list of the hardware components present in the Van). The 'Pan in the Van' BCC tools and training programmes use readily available information packages on the government schemes, technical guidelines, as well as pre-recorded contents in the form of messages, songs, games commentary, and training tips. This enables the 'Pan in the Van' team to focus on the tasks and give their full attention to the participants, reducing the risk of overlooking the steps during the training or distortion of the messages.

This approach also facilitates the planning and implementation of different NBA components, and supports the review and post-triggering in a synchronised manner. The 'Pan in the Van' has been incorporated into a larger WASH enterprise framework, which is set up at block levels equipped with a helpline (reachable through a toll-free number), a WASH store, a mobile unit (the 'Van') and a production centre, if

Table 14.1:
Mobile van components

Hardware Components of the Mobile Van	
• 12-seater AC van (with modular carriers and cupboards) • Folding canopy, display tables and tray, display boards, etc. • Dias, podium and tent (for 15–20 persons) • IEC materials (games, videos, plays, folk songs, games, recording of key messages, etc.)	• Audio-video equipment and public address system • Models of sanitation facilities (various options) • Equipment for producing sanitary items (pans, p-trap, etc.) • Generator for providing power backup • Basic furniture such as chairs, tables, carpets, etc. • Trophies and certificates to encourage participants

Figure 14.1:
'Pan in the Van' as part of a WASH enterprise framework

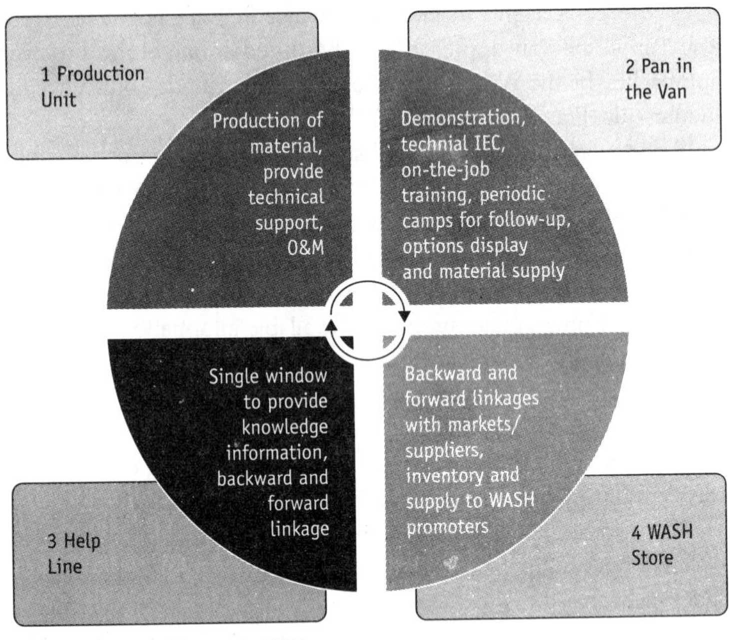

Source: EEDS own eleboration, 2013.

required (Figure 14.1). The WASH enterprise is embedded in a network of local WASH promoters and interventions in the specific *Gram Panchayats*. The WASH promoter is from the same village in which she/he operates, linked to a toll-free number telephonic helpline and facilitates the process to bring about the sanitation improvement in his/her community.

The 'Pan in the Van' related tools and methods offer opportunities for women and children to participate creatively and joyfully. Features such as the round-the-clock stay for three days of the 'Pan in the Van' team/ unit at the camp site (the flexible camp schedules, the on-site training facilities, exhibition of options, IEC tools, production machines, separate meeting arrangements and community reviews) enable different sections of society, especially women, to learn and contribute as per their own pace and according to their own needs. As part of its community and school interventions, the 'Pan in the Van' comprises dedicated tools and methods involving women to address their strategic needs. Through this, village-level institutions are strengthened, change agents emerge and follow-up mechanisms are set up. A *Mahila Toli* (women group) is formed to extend support in leading and spearheading the WASH interventions. The approach is based on a four-step process. First, a participatory hygiene analysis is undertaken to understand the local situation. Second, an on-site three-day camp is conducted to create community awareness, to build institutional capacity focusing on women, to enrol children for school sanitation and hygiene education (SSHE), and prepare the village (or *Panchayat*) for WASH plan. Third, the institutions are strengthened, village-based 'enterprises' are created and trained, and water security plans are prepared. Fourth, sustainability is ensured through regular community review in a participatory and joyful way.

The 'Pan in the Van' approach creates awareness generation through its mobility (the entire process being mounted into a van) and reaches hitherto unreached marginalised groups in both urban and rural areas in improving access to WASH services. It focuses on team building for local collective behaviour change: this includes the capacity development of key stakeholders such as the village water and sanitation committees, women self-help groups (SHGs), school functionaries, etc. The focus is also laid on strengthening the supply chain and enhancement of skills of service providers such as masons, plumbers, NGOs, etc. The mechanism for WASH action planning and community review includes the facilitation of action plan preparation to accelerate the elimination of open defecation in line with the TSC/NBA targets at the *Panchayat* level with clear roles and responsibilities. It also covers establishing and facilitation of a sustainable community-based monitoring system. This approach refrains from creating new/parallel institutions; instead, it focuses on developing the capacity of existing institutions/groups such as the local mandated development committees (i.e. village health/WASH-related committee), the schools, PRIs, SHGs, etc. It reduces the transmission losses and distortion of messages, as on-site capacity building is conducted to reach the target groups directly. Furthermore, the village-level WASH action plans are prepared in such a manner as to ensure that the implementation would not have to rely on additional financial resources. It promotes convergence to meet financial needs at the *Panchayat* level. Additionally, it institutes a simple, transparent and joyful format for regular community review.

14.3 Salient Women-centric Features of the 'Pan in the Van' Approach

The 'Pan in the Van' process intends to overcome some of the key barriers typically observed in the field that often prevents women from participating more widely in conventional sanitation and hygiene programmes. How this is done is outlined in Table 14.2.

Table 14.2:
Tested 'Pan in the Van' solutions to address key barriers for women's participating

S. No.	*Reasons for Low Participation of Women in Conventional Sanitation Training Programmes*		*How the 'Pan in the Van' Approach Seeks the Pro-active Participation of Women*
	Reason	*Detail*	*Tested Solution*
1	Lack of information	Information is typically passed on from man to man	Information is passed on to the woman members of the family by women 'Pan in the Van' team members as well as through the children enrolled in the 'Pan in the Van' games
2	Illiteracy	Women literacy in rural areas is lower than national average. Many rural women are unable to read descriptive hand-outs given in trainings	Involves pictures, videos, drawings and activities. Involves activities that can be undertaken within the village boundaries
3	Confined only to one or few aspects	Most approaches focus on a single or limited aspect such as triggering, supply or marketing, hand-washing and toilet construction	Offers a holistic palette of solutions related to WASH issues, covering all key components, including menstrual hygiene, construction, backward and forward linkages of supply chain management and services, as well as governance and institutional strenghtening in a coordinated manner
4	Addressing limited target groups	The processes and events typically target individual households and/ or involve specific target groups at one time	Focusses on the community as a whole, including school children and physically challenged persons with distinct strategies and methods simultaneously

(Table 14.2 Continued)

(Table 14.2 Continued)

S. No.	Reasons for Low Participation of Women in Conventional Sanitation Training Programmes		How the 'Pan in the Van' Approach Seeks the Pro-active Participation of Women
	Reason	Detail	Tested Solution
5	Exclusion due to off-site trainings or activities (venue far away from villages)	There are chances of exclusion, especially of women, as most of the conventional strategies use off-site activities by which inputs are provided (training of trainer's mode) to some selected community representatives mostly away from their place of living (often at block, town or district level) with the expectation that they would disseminate the skills and information to their community members upon returning to their village	The events and trainings are conducted locally, within the village. The entire village is enrolled as the events are conducted in the open where anybody can watch or participate. This considerably increases the chances of women participation. The equipment used is mobile (through the Van) and activities are performed where the community resides and where real challenges are found
6	Monotonous classroom-based theoretical training	Trainings modules are often given in classroom style. They mostly involve theoretical learning and ignore the varying levels of learning capacity of the participants	The tools and methods involving the community are simple and joyful. Audio-visual aids help building skills amongst community in a more interactive manner, amplifying the support provided by 'Pan in the Van' technicial team. Mascots are used as brand ambassadors of cleanliness and faecal matter
7	Scheduling	Rigid time schedule do not allow women and marginalised groups to participate	Through the on-site approach, the 'Pan in the Van' team stays and functions from within the village to encourage women and daily wage earners to participate as per their convenience. The schedues are customised for each camp as per the interest and lifestyle of women in that area

(Table 14.2 Continued)

(Table 14.2 Continued)

S. No.	*Reasons for Low Participation of Women in Conventional Sanitation Training Programmes*		*How the 'Pan in the Van' Approach Seeks the Pro-active Participation of Women*
	Reason	*Detail*	*Tested Solution*
8	Men-centric exercises	Most of the activities are confined to the awareness raising of men and low attention is typically given to women's participation	Activities do not require strength or any particular skills and are adapted from regular existing games, many of which are played by women and girls. The 'Pan in the Van' approach intentionally creates opportunities for the participation of women and empowers them to take leadership roles, to actively take part in local governance and become change agents. The processes are suitable even for the physically challenged
9	Men-centric issues	Many approaches address popular issues such as water for irrigation and the construction of latrines (by male masons), which are typically confined to and considered important for men	The topics and issues keep the concern and needs of women at their the centre, i.e. household chores such as fetching drinking water, washing utensils and food, maintaining personal hygiene, washing clothes, bathing, use of latrines, kitchen garden, livestock waste, etc. Women can follow the messages without any cultural ideology being hurt
10	Pre-conceived notions about the interest and capabilities of women	Pre-conceived notions about the interest and capabilities of women very often prevail in traditional sanitation programmes: examples of misconceptions include that women cannot be trained on technical aspects, they don't perform when it comes to planning and decision-making, etc.	Myths about various aspects are removed through discussion and practical demonstrations. Women are also encouraged and empowered to rise as 'sanitation champions' and 'natural leaders'

(Table 14.2 Continued)

(Table 14.2 Continued)

S. No.	*Reasons for Low Participation of Women in Conventional Sanitation Training Programmes*		*How the 'Pan in the Van' Approach Seeks the Pro-active Participation of Women*
	Reason	*Detail*	*Tested Solution*
11	Tokenism in participation	The participation is measured in terms of numbers without creating suitable opportunities or an enabling environment. The absence of women is taken as lack of interest on their part	A separate 'Pan in the Van' tent is set up to discuss women and adolescent girls' issues, including menstrual hygiene management, with female members of the 'Pan in the Van' team. Other specifically tailored activities involve women in each step of the development of the village action plan in view of rendering it open defecation free
12	Process is overlooked	Focus is often given solely on the inputs (funds, material, etc.) and outputs (number of toilets built); little is usually given to process aspects	The 'Pan in the Van' approach conducts training in a step-by-step manner and puts emphasis on follow-ups to sustain interest. It triggers and stimulates the community to become and remain involved

14.4 The 'Pan in the Van' Process

The four steps of the 'Pan in the Van' approach are conducted as per the representation in Figure 14.2, with further details provided in the subsequent text.

Step 1—Participatory health hygiene analysis and environment building: During a one-day event, key baseline information is collected to understand the background situation of the target groups, the local situation related to WASH in the light of the 'Pan in the Van' objectives, listed earlier. After the initial rapport building with local leaders and opinion makers, the team conducts primary and secondary information collection in a planned manner: the primary data collection is done using a standard checklist on a sample household basis as well as through interactions

Figure 14.2:
Conceptual representation of the four steps of 'Pan in the Van' approach

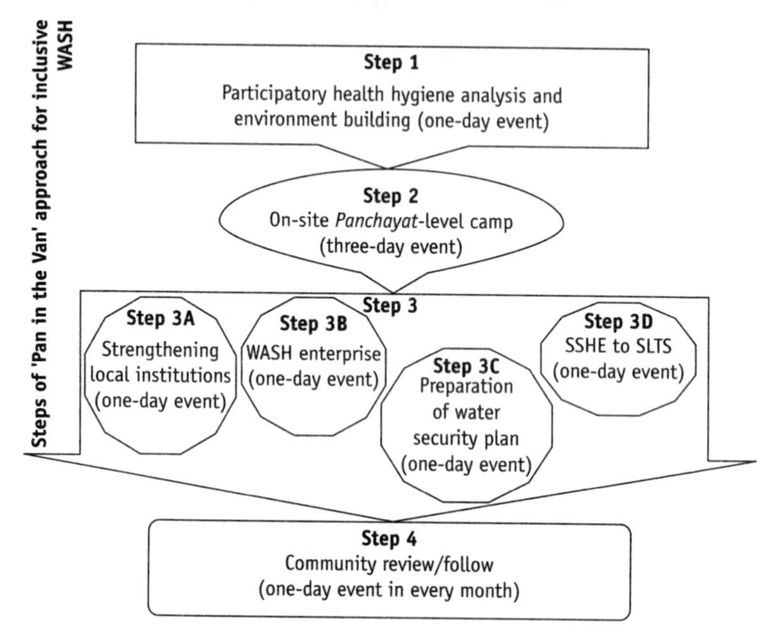

Source: EEDS own eleboration, 2013.
Note: All activities of Step 3 are one-day events, done in sequence.

with key stakeholders and frontline workers. The secondary data related to infrastructure and resources are collected from the *Panchayat*. A water quality expert and a medical doctor also carry out drinking water testing and heath check-ups of selected individuals. At the end of the day, the medical doctor facilitates the Focus Group Discussion (FGD) around the findings of the survey and initiates a reflection with the community members around the WASH-related issues in the village and/or *Gram Panchayat*. In the FGD key members of the community participate: women SHGs, the village-level water and sanitation committee, PRIs and other community members, with a special focus on children, youth, women and adolescent girls. The information collected is compiled in the form of a *Panchayat* status report on WASH and preliminary planning is made for Step 2, the *Panchayat*-level three-day camp.

Step 2—On-site* Panchayat*-level camp: During the second stage, 'Pan in the Van' team continuously stays in the *Panchayat* for three days and conducts a camp, based on a set of detailed daily schedule and clear objectives. A typical camp covers the creation of community awareness and the triggering for action around the WASH-related issues identified in Step 1. The three-day camp focuses on capacity development of key institutions such as the village-level water and sanitation committees, SHGs, masons and other relevant stakeholders. School children are also engaged through the introduction of SSHE [using school-led total sanitation (SLTS) methods], and WASH action plans are prepared, including a community review/monitoring system. An overview of the tools and games available and used as part of the 'Pan in the Van' toolbox during the three-day camp can be found in Annexure 14.1 at the end of this chapter.

Step 3—Strengthening local delivery mechanism and institutions: A few days after the three-day camp, the 'Pan in the Van' team returns to the same village/*Panchayat* for a follow-up session, in order to further build the capacity of different key stakeholders. They strengthen local institutions that are responsible for WASH activities and mainly consist of women 'natural leaders'. They also improve supply chain management, impart skills to masons, and orient the service providers and suppliers to ensure quality and timely services. SSHE in which children act as WASH messengers and agents of change to households and influence the community to adopt the four key hygiene practices (safe storage and handling of drinking water, use of toilet and safe disposal of child excreta, as well as hand-washing with soap at critical occasions during the day).

Step 4—Community review and follow-up mechanism: The *Panchayat*-level WASH plans prepared during Step 2 (three-day camp) are reviewed each month by the community. Follow-up facilitation is conducted by the 'Pan in the Van' team. To initiate the review process, some of software activities and games of the 'Pan in the Van' toolbox (see Annexure 14.1) are used to reinforce the key messages and also to continue keeping the community engaged. During the first 5–6 months, the 'Pan in the Van' team initially facilitates the review process; later the same is then entrusted to the *Panchayat* and the community themselves.

14.5 Key Results

Out of the four steps, the second step (three-day 'Pan in the Van' camp) was organised between 2009 and 2012 in 80 *Panchayats* across five districts of Madhya Pradesh with encouraging results, which are briefly described below.

1. *Entitlement realisation:* The approach not only triggered the demand for improved sanitation but also built the team and sustained the programme in a more inclusive manner. It benefited people in pursuing their entitlements, increasing the access to dignified sanitation and created space for excluded groups to be informed and seek active participation.
2. *Inputs for policy formulation:* Based on the learning and inputs from the 'Pan in the Van' approach, new dimensions were incorporated in the IEC Strategy document of the Government of Madhya Pradesh to accelerate the elimination of open defecation in Madhya Pradesh through its sanitation campaign. In 2012, the Government of Madhya Pradesh adopted the national NBA framework and tailored it to a women-centric sanitation campaign, using the principles of CLTS as its operating system.
3. *Coverage:* The camps improved the sanitation coverage in 120 villages of 80 *Gram Panchayats*. It also strengthened WASH-related services and capacity in nearly 100 *Anganwadis* (pre-school children development centres), about 240 schools and around 20,000 households. These camps provided opportunities for about 12,000 women and girls to actively learn and participate in the improvement of the sanitary status of their community, and also enrolled 100 *Anganwadi* workers, 700 school teachers and 35,000 school children to become agents of change in their communities. From the rapid assessment conducted, taking a sample size of 10 *Gram Panchayats* (EEDS, 2010), it is estimated that around 1,600 households gained access to improved WASH services and/or adopted key hygiene behaviours as a result of the 'Pan in the Van' interventions.
4. *Impacts:* A third-party rapid assessment of 10 of these 80 three-day camps was conducted (EEDS, 2010). The findings indicated that

many women reported that the three-day 'Pan in the Van' camp was the first experience and sound opportunity for them to explore their roles outside of the confinement of their households. They were able to use their newly gained ability to contribute more strategically to the village's WASH issues and to be recognised by men. Many women reported of having gathered first-hand experience on planning and analysis while preparing their village WASH plan. The team-building processes as well as the collective problem analysis and solving proved to provide women with a new set of skills to work in a collective manner to achieve common goals. A collective thought process was initiated by which people took interest in learning about the details of government schemes, entitlements and interrelation of behaviour, health and governance. Households started storing drinking water in a safer and more hygienic manner and the community started to approach the duty bearers to raise demands about their entitlements. Furthermore, the camps resulted in a sense of action and responsibility. Service providers were also provided new ways of reaching to the community. Around 600 masons were trained on different technology options, quality aspects and supply chain management for toilet construction in the remotest locations. Bottom-up monitoring systems based on transparency and fun. Finally, the approach provided space for participation as the tools promoted the inclusion of women through tailored and intentional opportunities. The target population could participate as per their schedule and availabilty without affecting their work and livelihood.

5. *Quantitative assessment:* A rapid assessment was made for 15 major areas rating them on a scale from 1 to 10 (with 10 as the highest), based on defined criteria (EEDS, 2010). The assesment results are presented in Figures 14.3–14.5. From Figure 14.3, it is interesting to note that the role of women in local-level governance scored well (6) and that the increase in their WASH-related awareness scored as high as 8. The analysis presented in Figure 14.4 underlines the positive impacts the 'Pan in the Van' 3-day camps had on women's participation and knowledge transfer.

The retention of the key messages delivered during the three-day 'Pan in the Van' camps, assessed six months after the camps

Figure 14.3:
Rating of main contributions of 'Pan in the Van' approach based on defined criteria

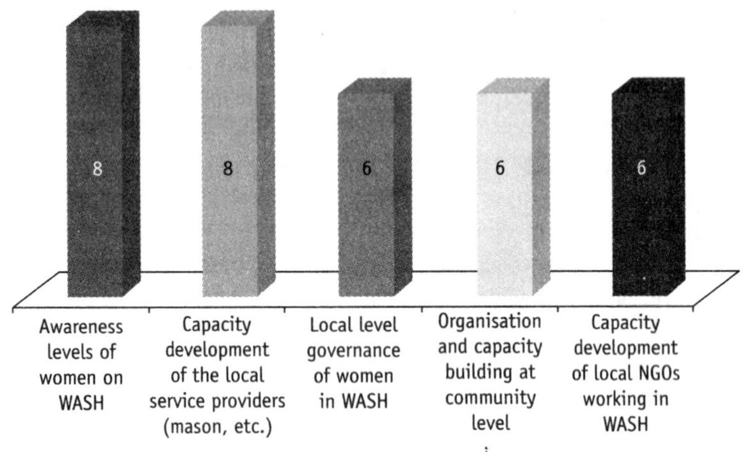

Source: EEDS (2010).

Figure 14.4:
Rating of the effectiveness of the 'Pan in the Van' approach with regard to given aspects

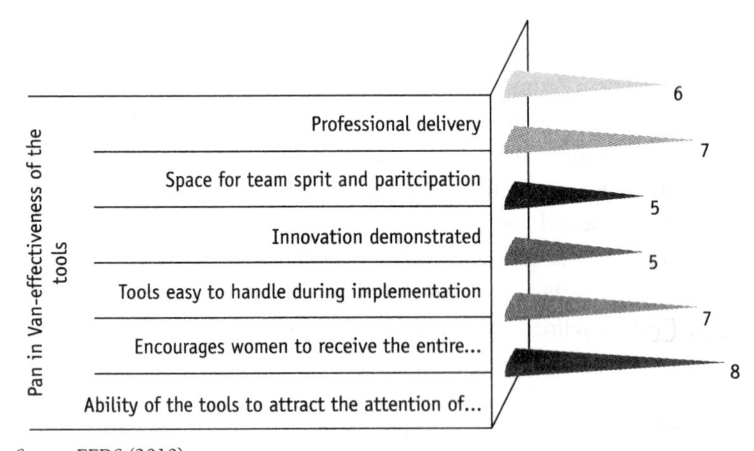

Source: EEDS (2010).

took place in 10 sample *Gram Panchayats* was over 60 per cent. Around 500 households were reported of having stopped practicing open defecation and making conscious efforts to sustain the key WASH habits inculcated during the 'Pan in the Van' camps,

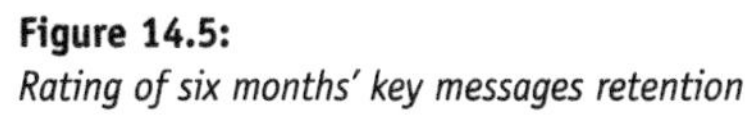

Figure 14.5:
Rating of six months' key messages retention

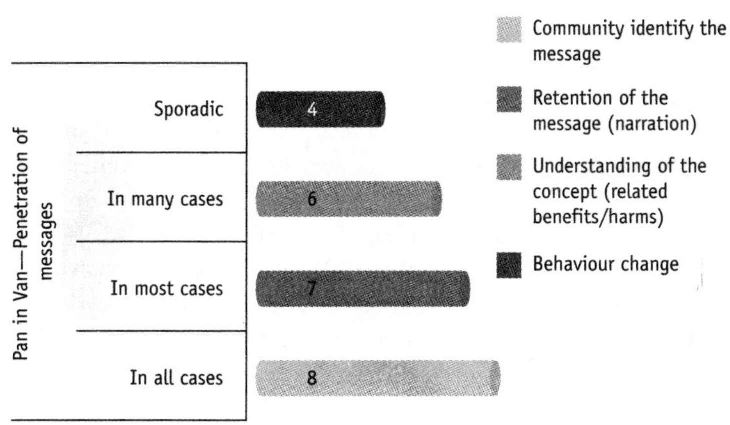

Source: EEDS (2010).

including the use of toilets, the use of soap for hand-washing at critical times, and the safe storage and handling of drinking water. Figure 14.5 presents the scoring overview of the message retention analysis.

6. *Demonstration of alternatives and technical aspects:* Finally, the display during the camps were reported to have provided rare opportunities for women to understand the technical aspects and options in sanitation such as leach pit method, p-trap, pan production, EcoSan, sub-community toilets, NADEF, vermi-composting and biofertilisers, as well as water purification.

14.6 Conclusion

This chapter has presented the main features of 'Pan in the Van' as an innovative approach for inclusive WASH. We have shared some findings of the first round of field testing, and its women-centric approach to tackle the pressing WASH issues, especially related to the elimination of open defecation. The chapter outlined the four steps of 'Pan in the Van' for capacity building through different innovative BCC tools as well as various methods to actively involve and empower women in a joyful manner around key WASH issues. The approach has been developed keeping women at the core of the entire process, suitable to their context and lifestyles and empowering them to take leadership roles. It offered new ways to show how women and children are enabled to become fundamental to sanitation improvement. The chapter also shed light on the fact that capacity building is a process whereby a community equips itself to undertake the necessary functions of governance and service provision in a sustainable manner. The chapter then drew from the main findings and lessons learnt from 80 camps in five districts of Madhya Pradesh, between 2009 and 2012. The initial evaluation showed how women and children could be an integral part of the development process and can take up leadership roles. The third-party rapid assessment showed improvements in message retention, behavioural change, empowerment of women through specific WASH village planning and the active participation of children.

Although the 'Pan in the Van' approach has been developed specifically to address WASH-related issues, its mobile and on-site capacity-building attributes as well as its women-centric principles can be easily transferred and applied to other sectors, such as health, nutrition and eduction. These attributes can be used to fill the capacity gap of other key development sectors, which typically suffer from similar gender and geographical inequalities. The 'Pan in the Van' approach has already started exploring the integration between its WASH applications and their linkages to health impacts by incorporating the services of a medical doctor in its team.

The operational cost of the four-step 'Pan in the Van' approach is estimated to be around ₹250–300 per household, excluding one-time fixed costs. The approach can therefore be considered to be cost-effective as compared to conventional capacity building and training methods. The logistical expenses are minimal and no travel expenses for the village participants are required in this on-site approach. The field applications of the 'Pan in the Van' approach have demonstrated its ability, and indeed comparative advantage, in reaching regions where most of the government schemes and services are less likely to penetrate. The 'Pan in the Van' approach thus offers a strategic opportunity to address inequities and disparities in the WASH sector, by purposefully delivering services to the most deprived communities.

Disclaimer

The views expressed herein are those of the authors and do not necessarily reflect the views of UNICEF or the United Nations.

Acknowledgement

The authors would like to recognise the valuable inputs provided by Dr S. Ramesh Sakthivel of IIT, Delhi and Ms Ashu Saxena from EEDS during the development and improvement of the Pan in the Van approach, and they are grateful to Dr Aidan A. Cronin, UNICEF Delhi, for his suggestions in improving an earlier version of the present chapter. We are grateful to the *Zila Panchayats* for assisting EEDS in conducting the camps in five districts of Madhya Pradesh. The assistance received from the communities, PRI representatives and government officials of Madhya Pradesh were crucial for taking this approach forward, and the authors would like to thank them on this occasion.

Notes

1. See the following link for further information: http://www.solutionsforwater.org/wp-content/uploads/2011/11/PAN-IN-THE-VAN-APPROACH-eeds-Bhopal-Nov-2011-2.pdf (accessed on 7 August 2013).
2. See the following link for further information: http://www.solutionsforwater.org/solutions/pan-in-the-van-%E2%80%93-an-innovative-on-site-approach-for-inclusive-wash (accessed on 7 August 2013).

References

Chambers, Robert and Gregor von Medeazza. 2013. Sanitation and stunting in India: Undernutrition's blind spot. *Economic and Political Weekly*, 48(2): 15–18.

EEDS. 2010. End-term project evaluation, done by Mr P. Mandal (independent consultant), Energy, Environment and Development Society (EEDS), Bhopal.

EPW. 2012. Toilets can be temples. *Economic and Political Weekly*, 47(42): 7.

Government of India. 2012. Census of India 2011. Houses, Household Amenities and Assets. Latrine Facility. Government of India.

Liu Li, Johnson L. Hope, Cousens Simon, Perin Jamie, Scott Susana, Lawn E. Joy, Rudan Igor, Campbell Harry, Cibulskis Richard, Li Mengying, Mathers Colin and Black E. Robert. 2012. Global, regional, and national causes of child mortality: An updated systematic analysis for 2010 with time trends since 2000. *The Lancet*, 379: 2151–61 (for the Child Health Epidemiology Reference Group of WHO and UNICEF).

UNICEF. 2012. Equity in drinking water and sanitation in India. Perspectives on equity and gender in the WASH Sector in India. UNICEF, New Delhi, India.

UNICEF, FAO and SaciWATERs. 2013. *Water in India: Situation and Prospects*. New Delhi, India.

WSP. 2011. Economic impacts of inadequate sanitation in India. Water and Sanitation Program, World Bank.

Hutton, Guy and Laurance Haller. 2004. Evaluation of the costs and benefits of water and sanitation improvements at the global level. World Health Organization, Geneva.

WHO and UNICEF. 2013. Progress on sanitation and drinking water: 2013 update. WHO/UNICEF Joint Monitoring Programme for Water Supply and Sanitation, Geneva/New York.

Annexure 14.1:
'Pan in the Van' toolbox

S. No.	Tool	Objective	Details	Participants
1	*Langdi* game	To sensitise the community about open defecation	This is based on a popular rural game played by girls and women. Participants make their village map (using rangoli colours, powder, etc.) indicating the main places related to WASH, including the open defecation sites. Once the map is drawn, the participants are asked to cross the village map without touching the dirty places and open defecation points.	Women and girls
2	Ring game	To raise interest and communicate key WASH messages	The popular ring game is used to spread and reinforce the WASH messages. Rings are thrown by participants on a chart showing *do's* and *don'ts*. Participants explain the issue where the ring lands.	Women and children
3	Chair game	To raise interest and communicate key WASH messages	The well-known musical chair game is played with 15 chairs representing key WASH messages. The participant occupying the chair explains the particular message.	Women and children
4	Songs	To raise interest and communicate key WASH messages	Based on a series of popular Bollywood songs, key WASH messages are raised in an entertaining and creative way. These songs continue to be played throughout the duration of the camp.	Community
5	*Mallasure ki Kahani* (story)	Scientific WASH facts packed in a story form to sensitise the community	A role play performed on stage presents, with the help of banners and mascots, the key WASH messages to the audience. The villagers themselves also participate in the role play. The mascots play as WASH brand ambassador.	Community
6	PRA	To understand the WASH situation of the village	The PRA concept customised for WASH situation analysis and planning.	Community, Women

(Annexure 14.1 Continued)

(Annexure 14.1 Continued)

S. No.	Tool	Objective	Details	Participants
7	*Dandi Yatra* (village transect walk)	To identify clean/ dirty areas in and around the village	Rallies are conducted by students with banners and posters around WASH-related issues. They visit various places significant for hygiene and sanitation. The resource persons explain and demonstrate various aspects.	Children
8	Joker game	Orientation on WASH-related message in joyful manner	Mascot and brand ambassador of cleanliness and hygiene. It roams around in the village, raising attention and spreading WASH messages in a fun manner.	Community
9	School kit	To learn and monitor WASH services inside the school premises	Provided to the schools in the form of study material and a monitoring tool.	Children
10	Pan production 'Produce your PAN'	Demonstration of low-cost options	Demonstration of options with the basic technical details, where anyone can bring their own raw material (often from locally available recyclable waste material, such as brick dust, etc.) and prepare their own sanitary pan.	Masons and community
11	Football match (*Kick to Kitanu*)	Monitoring and environment creation	Children (and sometimes also adults) play a football match between the teams of 'Swachhata' (cleanliness) and 'Kitanu' (bacteria). The commentator broadcasts the WASH messages along with the game events.	Children, youth and community
12	Videos and Movies	Awareness generation	Showing the message to the entire community through interesting films during the evening, in the village main gathering place.	Community

15

Liberty from Shame: Accelerating Sanitation with ASHAs

Amit Mehrotra and Ajay Singh

15.1 Introduction

Questions of needs, interests, access and control of resources and services are based on a variety of factors, including gender. An integrated approach to water and sanitation recognises these differences and the disparate priorities they create for women and men. The involvement of women and girls is crucial to effective water and sanitation projects. Women and girls in developing countries bear most of the burden of carrying, using and protecting water. They are also mostly responsible for environmental sanitation and home health. Given the present roles of women in water and sanitation, their active involvement and empowerment is needed for water and sanitation efforts to be successful, and without further adding to their burden. Gender mainstreaming is needed to achieve gender balance and reduce inequalities suffered by women and girls.

Open defecation is a prevalent cultural behaviour in most parts of rural India. This, along with the relative neglect of sanitation in terms of development priorities, was reflected in the country's low sanitation coverage at the close of the 1990 with only 7 per cent of the rural population using improved sanitation facilities (WHO/UNICEF, 2012). It was found that only one in five rural households had access to a toilet (Census of India, 2001a). With the additional low awareness of improved hygiene behaviour, achieving the goal of total sanitation became a pressing challenge in rural India. In response, the Government of India launched the Total Sanitation Campaign (TSC) in 1999 with the goal of achieving universal rural sanitation coverage by 2012. The responsibility for delivering on programme goals rests with local governments (*Panchayati Raj* Institutions—PRIs) with significant involvement of communities. The TSC advocated a shift from a high- to low-subsidy regime, greater community involvement, demand responsiveness, and the promotion of a range of simple and cost-effective latrine options. The state and central governments had a facilitating role that took the form of framing enabling policies, providing financial and capacity-building support, and monitoring progress (Ministry of Drinking Water and Sanitation).

Uttar Pradesh, one of the largest populated states of India, has been one of the key states to embrace TSC. Since 1 April 2012, this campaign is being referred to as *Nirmal Bharat Abhiyan* (NBA). It has made concentrated efforts to reach each and every household in the rural areas, but has failed to achieve the desired results. The 2011 census figures showed a relatively low progress of sanitation coverage in Uttar Pradesh; only 22 per cent households were having access to latrines. With this pace, it will take another 33 years to reach the Millennium Development Goal targets (as per the current trend line, it will achieve only 24 per cent by 2015) and another 78 years to claim Open Defecation Free (ODF) status. The figures on coverage related to sanitation in Uttar Pradesh are disturbing. Compared to all-India figures, only 2.6 per cent of people gained access to latrine within premises in rural Uttar Pradesh during the decade. There was marginal increase in rural households having latrine from 2001 to 2011 (19.9. per cent to 21.8 per cent), whereas, during the same period, 8.8 per cent households in rural India gained access to latrine within premises (Census of India, 2011b; see Figure 15.1). However, the online reporting of the sanitation programme presents a very different picture

Figure 15.1:
Progress in improved sanitation, Uttar Pradesh (Rural)

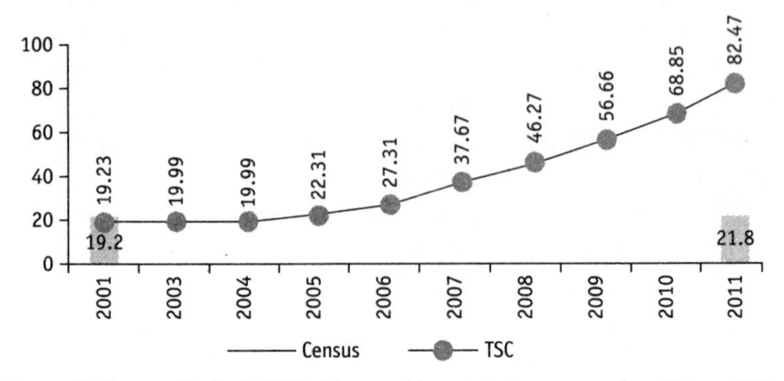

Source: (1) Census of India (2011b), (2) www.ddws.nic.in/tsc—accessed on 23 June 2012.

of the coverage during the same period. The graphs clearly indicate that there is a gap of more than 60 per cent between the two sources.

One of the key issues for such low coverage is lack of participation of different stakeholders in the sanitation programme. Women, one of the major driving forces to adopt sanitation practices, were not involved in the decision-making process at the local level and their participation was minimal. Women play a vital role in raising awareness about sanitation issues in their communities, and improved water and sanitation sources are the first step towards empowering women.

This chapter narrates the experience showing improvement in household sanitation practices with the help of interpersonal communication, spearheaded by ASHA. The chapter also outlines the experience of UNICEF-supported initiative in the eight districts of Uttar Pradesh. It also highlights the various approaches involving capacity building of ASHAs and group-based communication for strengthened counselling by ASHA leading to improved WASH practices at the household level. The chapter outlines the case studies elaborating the role of ASHA in promoting WASH practices.

15.2 Objective of Sanitation Programme

TSC is a comprehensive programme to ensure sanitation facilities in rural areas with the broader goal to eradicate the practice of open defecation. TSC as a part of reform principles was initiated in 1999 when Central Rural Sanitation Programme was restructured making it demand-driven and people-centred. It followed a principle of 'low to no subsidy' where a nominal subsidy in the form of incentive is given to rural poor households for construction of toilets. The strategy was to make the programme 'community-led' and 'people-centred'. A 'demand-driven approach' is to be adopted with increased emphasis on awareness creation and demand generation for sanitary facilities in houses, schools and for cleaner environment (Ministry of Drinking Water and Sanitation).

TSC also emphasised the idea of involving women, particularly those who are either elected representatives or are working as grassroots functionaries in the villages to promote sanitation and hygiene practices. This clearly paved the way for intensively roping in ASHA, *Anganwadi* workers, etc., to be the change agents and motivate other women of the village to demand sanitation facilities at the household level and help the community to achieve better living conditions.

15.3 UNICEF's Eight District Approach

Realising the need of including women and recognising the importance of sanitation, UNICEF's Uttar Pradesh office strategically started advocating for improving services by village level grass root workers through social mobilisation processes. ASHA being one of the key health functionaries, having defined role to promote sanitation and health in the village, became the natural choice. During the starting phase, it was found that although ASHAs organised village-level meetings and imparted knowledge regarding sanitation, they still needed orientation and skills to be upgraded to make them as sanitation motivators. This included equipping them with resource materials and offering trainings so that they were more conversant with hygiene and sanitation issues, as well as had the desired level of expertise to address sanitation issues. The strategy of capacity building of ASHA started from Lalitpur district where all ASHA workers were trained in phases. Before imparting training, modules were prepared and state-level master trainers were trained. As of now, there are 694 of the ASHAs trained in Lalitpur district alone. These ASHAs then included the sanitation message in their programme in Village and Health Nutrition Days (VHND). From this the intervention spread to other districts and gradually capacity-building interventions were taken up in seven more districts, which are UNICEF focus districts in the state of Uttar Pradesh (see Figure 15.2). The highlighted districts were Moradabad, Lalitpur, Allahabad, Jaunpur, Varanasi, Mirzapur, Sonbhadra and Sant Ravidas Nagar (Bhadohi).

15.3.1 Strategic Focus-Capacity Building

The first strategy of the programme was to develop appropriate resource materials, which were to be used during community mobilisation by the ASHAs. A set of pictorial booklets was developed with messages in local

Figure 15.2:
Map of eight districts covered by UNICEF (2011)

Source: UNICEF (2011).
Note: This map is not to scale and does not depict authentic boundaries.

language, which the ASHAs carry with them during their group meetings and home visits. This helped them a lot in making people understand the importance of hygiene and sanitation. ASHAs were particularly oriented to speak more about the benefits of using toilet at home and safely disposing children's excreta. They emphasised the fact that these behaviours can save the children from diseases like diarrhoea, which is one of the biggest killers in rural areas. The health benefits promoted by ASHA are in a way extension of their own work and also helped them build strong rapport with the community.

Capacity-building programmes for ASHA workers played a major role in making them a perfect change agent as well as in accelerating sanitation in the villages. Table 15.1 shows the number of ASHAs who have been capacitated on WASH issues in UNICEF focus districts.

Table 15.1:
Training status of ASHAs on WASH issues in focus districts

District	Total Number of ASHAs	Total Number of ASHAs Trained on WASH	In Per Cent
Allahabad	3,865	2,885	74.6
Jaunpur	4,118	2,333	56.7
Lalitpur	806	693	86.0
Mirzapur	1,811	1,540	85.0
Moradabad	2,598	2,442	94.0
Sant Ravidas Nagar	1,233	1,067	86.5
Sonbhadra	1,477	1,005	68.0
Varanasi	1,956	1,314	67.2
Total	17,864	13,279	74.3

Source: Monthly Progress Reports of *Panchayati Raj* Department, Government of Uttar Pradesh, September 2012.

15.3.2 Promoting WASH in Village Health and Nutrition Days

Utilising the services of ASHA during the Village Health and Nutrition Day was another strategy to promote sanitation in the community. About 13,279 ASHA staff were trained between 2011 and 2012, and 1,748 VHNDs were organised in eight districts. Here, health and hygiene messages were provided to 20,069 women and 19,031 children. Simultaneously, the state Health Department was also taken into confidence to issue suggestive guidelines for ASHAs to participate in the programme. This was an important strategy to streamline the involvement and to scale it up to reach all the ASHA workers in the state. The programme now also recognises the importance of including ASHAs in promoting sanitation and they are an integral part of the implementation of NBA.

15.3.3 Demonstration Effect

One of the important aspects of the strategy was first to motivate these ASHA workers, along with other frontline functionaries, to adopt and

Table 15.2:
List of frontline functionaries having household toilets, Chajjalet, Moradabad

S. No.	Frontline Functionary	Total Number	Number of Officials with Latrine at Home
1	*Gram Pradhan*	77	77
2	*Panchayat* Secretary	14	14
3	Teachers	392	392
4	*Shiksha Mitra*	208	208
5	*Rozgar Sevak*	73	73
6	Auxiliary Midwife	34	34
7	ASHA	139	139
8	*Anganwadi* Teacher	193	193
9	*Anganwadi Sahayika*	194	194
	Total	1,324	1,324

Source: Monthly Progress Reports of *Panchayati Raj* Department, Government of Uttar Pradesh, September 2012.

construct latrines at their own households. This was meant to help them to demonstrate the benefits of latrines to other villagers. This strategy had a lot of positive feedback. People started listening to them as someone who is actually a user. ASHAs also found them at a better position to relate to the benefits they are talking about. It also had a cascading effect and in one development block (Chajjalet) of Moradabad district, not only ASHAs but also all the village-level frontline functionaries now have latrine at home (see Table 15.2).

The training programme has helped ASHAs in taking the sanitation programme at scale. Figure 15.3 clearly shows that the use of toilet is closely correlating the number of trained ASHAs.

15.4 ASHA as a Role Model: A Case Study

The following is the case study of an ASHA, Ms Reesa Maurya, who is now regarded as a champion for the cause of sanitation and is a role model for the entire district. Her story summarises the efforts made by her to make her village ODF.

Figure 15.3:
Trained ASHAs vs. toilet use in UNICEF focus districts

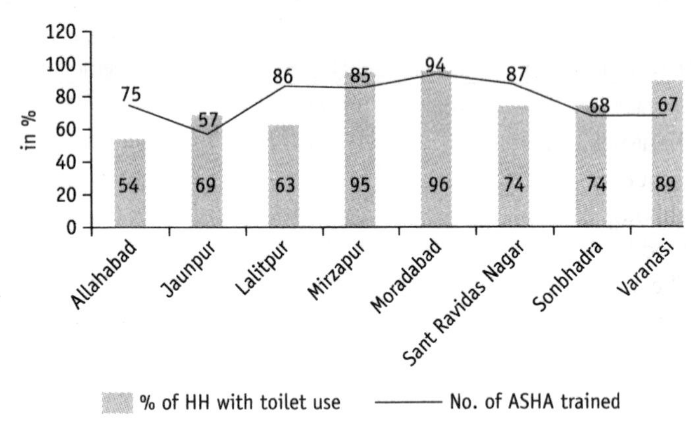

Source: Concurrent monitoring website of *Panchayati Raj* Department, Government of Uttar Pradesh, as on 13 March 2013.

Jagdishpur *Gram Panchayat* (village) of block Karanjakala is in Jaunpur district of Uttar Pradesh, India. Situated on the main road to Jaunpur, this village had a population of 779 inhabitants in 2011. Most families are engaged in commercial agriculture, growing vegetables and selling them to the urban areas of Jaunpur and are living above poverty line.

The case is important because it demonstrates how the training programme helped Ms Reesa Maurya to acquire the knowledge on WASH practices and this brought change in her behaviour and in turn made her a crusader for the cause of sanitation. Jagdishpur *Gram Panchayat* has an agrarian setup, and it was a general practice to use fields for defecation. No one thought of it as a problem. Reesa Maurya, being educated, got the chance to work as an ASHA. As part of her work, she got a chance to come in contact with women and discuss health issues as well as social issues. She realised that practically all the women faced problem of going out in the field for defecation. For them it was next to hell, a sheer shameful act they had to perform every day. Ms Reesa got a chance to attend the WASH training programme, which was organised by the *Panchayati Raj* Department and UNICEF with the support of District Sanitation Committee of Jaunpur. This acted as a turning point for her.

She came to know about TSC and how the lack of sanitation affected the health of children.

'During the two days of training on hygiene and sanitation, I got a good understanding of the negative impacts of open defecation that has disgusting consequences and creates an unhealthy environment and expenses for health treatment,' expresses Ms Reesa. In the past, I did not know the consequences of defecating in open. It was simply my habit like other neighbours in my village and we were not educated on the importance of good hygiene and technicalities of latrine construction.

She made it a point that after the training she would construct a toilet at her home and she did exactly that. She motivated her husband and other family members to construct and use it. 'I am very excited to have a latrine of my own,' Ms Reesa said. 'All my family members have started using the latrine. They drink safe water as well as clean the surroundings around the home.' Elaborating on her work, Ms Reesa adds, 'It is really a tremendous success for my family that at last we had built a hygienic latrine. It not only protects us from diseases but also enhances my family's social dignity. Now I think how my life has changed with just a few pieces of information. I feel very proud to have saved my child from excreta and water-borne diseases.'

What she was doing at home, she wanted the villagers to follow. So, she started organising group meetings with the women delivering the message of sanitation and hygiene. Her efforts slowly gained momentum. The same villagers who used to laugh at her started respecting her, and she, who once thought of leaving ASHA, started enjoying the reputation of someone who is intelligent and whose views are respected. Her effort has resulted in construction of latrines in at least 25 households, and she is now focusing on ensuring that the latrines whenever constructed are used.

'I wanted to prevent other people in my community from getting sick. I started asking all families in my village to start constructing latrines for their use and then our village will achieve the open defecation free status and bring good health for everyone, especially our children.' She added, 'Now we are all using hygienic latrines, washing our hands with soap after defecating. I am committed to working against open defecation.'

Now she is a role model for other ASHAs working in neighbouring *Gram Panchayats*. For *Panchayati Raj* Department of the district, she is a catalyst whose success is being replicated in other villages.

15.5 Results

Individual sanitation facilities have led to better access and improved personal hygiene, especially among rural women. For women and older girls, in particular, having a toilet at home meant privacy, which saved them from the dangers of going out before dawn or after dark for defecation, a practice that has serious side effects. Waiting so long to defecate leads to increased chances for urinary tract infections, chronic constipation and psychological stress. The biggest impact of the involvement of ASHAs has been that it has motivated large segments of the community to seriously ponder over the benefits of having individual sanitation facilities. They were instrumental in organising at least one VHND wherein ASHA discussed about WASH practices. With 13,279 people trained under ASHAs in the field, during years 2011 and 2012, 1,748 VHNDs were organised in eight districts.

In Uttar Pradesh, ASHAs are playing a big role, along with *Anganwadi* workers and other community-based organisations, such as SHGs, to accelerate sanitation coverage and supporting the strategy of making the villages ODF. Over a journey of 2 years, as of now more than 150 villages have become ODF in the eight districts, where UNICEF worked more intensively with the *Panchayati Raj* Department. Achieving ODF is no small task. It requires a lot of persistent efforts and involves the whole community to participate actively.

Around the world, achieving total sanitation in communities has proved to be an ongoing challenge for sanitation stakeholders. It requires whole communities to commit to stop defecating in the open and hygienically contain all faecal matter. The community-based approach adopted in the identified areas by UNICEF and *Panchayati Raj* Department ensured that the whole community is being engaged creatively to address poor sanitation and hygiene. ASHA being the catalyst helped in bringing the issue to the centre stage, as she was able to relate the problems suffered primarily by women. Leading women became the success mantra of the whole programme. They were able to highlight and demonstrate that having toilet is not only a necessity but also a pride for the household.

ASHA played an important role in making the community understand that poor sanitation affects everyone, and a collective approach is required to make the community ODF. She also played a vital role in facilitating the mobilisation of communities for collective action. She emerged as a 'natural leader' to facilitate community engagement and empowered the community to take decisions related to sanitation and hygiene in the village. The result of capacity-building support to ASHA has helped them in improving counselling and motivating people on household sanitation practices. In the eight districts, 77.26 per cent ASHAs got trained and the households using toilet also reached 76.60 per cent, which shows the high degree of correlation.

15.6 Conclusion

Despite significant investments in the last 20 years, India faces daunting challenges in the area of sanitation. The need of the hour is to recognise the importance of involving more number of women, be it a worker like ASHA or *Anganwadi* or community-based groups, to accelerate sanitation and make a healthy rural India. These women can function as strong catalysts of social change for achieving ODF.

There is a strong need to demonstrate convergence of resources, be it manpower such as an ASHA from Health Department or a village-level worker such as an *Anganwadi* worker, to make an integrated approach towards addressing the issue of sanitation and hygiene in rural India. Working in isolation may demonstrate sporadic incidences of success but it will never be able to take along all the members of the community together. The case of Reesa Maurya, an ASHA worker, clearly suggests that the issue of sanitation is very closely linked with health and if resources of health department are effectively utilised, it can demonstrate better results. ASHA is a health activist in the community whose primary responsibility is to create awareness on health and its social determinants and to mobilise the community towards local health planning and increased utilisation and accountability of the existing health services. She is essentially a promoter of good health practices in the villages. What Reesa Maurya has demonstrated, after intensive training programme, is something that needs to be institutionalised. ASHA's training should become an integral part of the overall strategy and rural areas get benefitted from it.

Involvement of village-level workers, especially women workers like ASHA also support in breaking the silence related to issues of gender in water and sanitation. In a country like India, we find that gender-specific roles are so deep that bringing parity between men and women, and placing them on the same platform are very challenging. Women and men usually have very different roles in water and sanitation activities; these differences are particularly pronounced in rural areas. Women are most often the users, providers and managers of water in rural households and are the guardians of household hygiene. If a water system breaks down, women, not men, will likely be the ones most affected, for they may have to travel further for water or use other means to meet the household's water and sanitation needs. Women are acutely affected by the absence of sanitary latrines:

- When women have to wait until dark to defecate and urinate in the open they tend to drink less during the day, resulting in all kinds of health problems such as urinary tract infections.
- Women are sexually assaulted or attacked when they go into the open to defecate and urinate.
- Hygienic conditions are often poor at public defecation areas, leading to worms and other waterborne diseases.
- Girls, particularly after puberty, miss school due to lack of proper sanitary facilities (Gender and Water Alliance).

A focus on gender differences is of particular importance, and gender-balanced approaches should be encouraged in plans and structures for implementation. Access to adequate and sanitary latrines is a matter of security, privacy and human dignity, particularly for women. ASHA, being a woman can ably spearhead this challenge of taking the message to community, motivating women members to demand for better and sustainable sanitation facilities at home and work towards changing the social norms related to gender inequality.

The involvement of ASHA in the dissemination of knowledge and taking lead in motivating community is the result of inter-sectoral convergence between sanitation programme and health (National Rural Health Mission). Such convergent strategy must also be replicated to bring in more partners. This will provide a renewed thrust and energy required to address sanitation in a holistic manner. Similarly, there is also the need to develop gender specific communication strategy addressing sanitation and the important role women play related to water and sanitation.

Traditionally communication approaches have largely tended to focus on women as homemakers and caregivers of children, the sick and elderly. Putting women at the centre of water and sanitation improvements will certainly lead to better health for all because:

- Women have a good knowledge about local water and sanitation practices and any associated problems, which can direct interventions.
- Women's interest in the family's health motivates them to bring about improvements.
- Women use group activities to reach other women and disseminate messages about good hygiene.
- Women are targeting men for involvement in sanitation and hygiene promotion so that they too take responsibility for this aspect of personal and family living.

At the community level, hygiene and sanitation are considered women's issues, but they impact both genders. Yet, societal barriers continually restrict women's involvement in decisions regarding sanitation improvement programmes. Thus, it is important that sanitation and hygiene promotion and education are perceived as a concern of women, men and children and not only of women. Separate communication channels, materials and approaches have to be developed to reach out to men and boys. It is also important to target community leaders for gender sensitisation; this would facilitate mainstreaming gender in sanitation and hygiene promotional activities.

Disclaimer

The views expressed herein are those of the authors and do not necessarily reflect the views of UNICEF or the United Nations.

References

Census of India. 2001a. Factsheet 'availability and type of latrine facilities'. 2 pages. Government of India, New Delhi.

———. 2011b. Datasheet of Uttar Pradesh, Available at http://www.censusindia.gov.in/ (http://www.censusindia.gov.in/2011census/hlo/Data_sheet/Uttar_Pradesh/rgi_up_datasheet_02.pdf (accessed on 31 March 2013).

Gender and Water Alliance (GWA). 2006. *For Her—It's the Big Issue: Putting Women at the Center of Water Supply, Sanitation and Hygiene*. Geneva: WSSCC, 36 pages.

Ministry of Drinking Water and Sanitation, Nirmal Bharat Abhiyan Guidelines, Available at http://www.mdws.gov.in/(http://www.mdws.gov.in/sites/upload_files/ddws/files/pdf/Final%20Guidelines%20%28English%29.pdf (accessed on 31 March 2013).

Monthly Progress Reports of *Panchayati Raj* Department, Government of Uttar Pradesh, 30 September 2012.

Panchayati Raj Department, Government of Uttar Pradesh, Concurrent monitoring, Available at www.st12.in (accessed on 31 March 2013).

UNICEF. 2011, 2012. Map of Uttar Pradesh.

WHO/UNICEF. 2012. Joint Monitoring Programme for Water Supply and Sanitation. ISBN: 978-92-806-4632-0. 66 pages. Geneva. Available at http://www.wssinfo.org/.

SECTION 4

Conclusion

16

Conclusions and Way Forward

Aidan A. Cronin, Anjal Prakash and Pradeep K. Mehta

16.1 Reflections

Despite tremendous progress, India still faces unprecedented water, sanitation and hygiene (WASH) challenges (UNICEF, FAO and SaciWATERs, 2013). Sustainability and empowerment are not, as yet, sufficiently addressed. However, communication, behaviour change and inclusive participatory processes are increasingly being seen as key to real change through impacting on those who feel the burden most—minority social groupings and above all, women (MDWS, 2012).

In the introduction to this collection, we argued that there is a lack of sufficient WASH and gender documentation in India. While excellent analyses for the broad water sector in India exist (e.g. Kulkarni, 2011; Zwarteveen et al., 2012), this book focuses primarily on the specific components dealing with drinking water and sanitation components in India with hygiene still not sufficiently reflected. All the chapters of this book show conclusively that the burden of drinking water and sanitation invariably is heavier on women, often at the expense of their health, education, income earning opportunities, and social, cultural and political involvement. But more importantly, the chapters highlight key opportunities that women have as change agents in their communities.

A key learning has been that a robust participatory process is vital. Lala et al. (Chapter 2) present timelines and tools in this respect to help ensure gendered outcomes in WASH programmes but such processes require detailed field application and requisite time to be successful. Kabir et al. (Chapter 3) and Chakma et al. (Chapter 9) re-emphasise this point in terms of underlining the key role of data and planning, but these approaches need to be eventually embedded in government programmes for scale and sustainability. Wani et al. (Chapter 6) detail several drivers of success for improving benefits to women and vulnerable groups. However, if India is to act on these best practices, much more capacity is required: Prakash and Goodrich (Chapter 4) and Sinha (Chapter 5). Mani et al. (Chapter 10) as well as Kale and Zade (Chapter 12) show clearly that such capacity building leads to success. Saxena et al. (Chapter 14) stress that such capacity building may need to be organised taking into account the timing and needs of women to ensure their participation and follow-up.

As per the National Rural Drinking Water Program (NRDWP) guidelines, the members in VWSC should be selected to represent various groups of society and 50 per cent of which should be women especially those belonging to SCs, STs and OBCs (MDWS, 2013). This has the potential to have a major impact on gender empowerment but enforcement is often lacking. As Joshi (2005: 145) noted, 'there is no project indicator to measure the effectiveness of Dalit representation in VWSC or to assess gender in Dalit representation.' Bastola (Chapter 7) echoes this point while reflecting on the point that when required participatory process is circumvented, then results will not be sustained. The MDWS Strategic Plan (MDWS, 2011) reiterates the need for an increased focus on gender in water and sanitation programmes but operational guidance, action and sustained follow-up are needed to achieve this.

The Special Rapporteur on the human right to safe drinking water and sanitation emphasises that states cannot fully realise the human rights to water and sanitation without addressing stigma (including gender) as a root cause of discrimination and other human rights violations (UN, 2012). A rights-based approach is required and power dynamics must be better understood to ensure equity and gender gains. Convergence of efforts of all actors is essential to achieve this, as Mehrota and Singh outline in Chapter 15.

The collection presents many success stories: Mehta and Saxena (Chapter 8) and Prasad et al. (Chapter 11) show that gendered outcomes and associated impact are possible even in the most impoverished and difficult environments. Such successes need strong monitoring and follow-up to ensure sustainability, as argued by Saxena et al. in Chapter 14 and Medeazza et al. in Chapter 13. Indeed the latter shows that sanitation success is possible with gendered outcomes.

Therefore, the experiences in this collection repeatedly emphasise that programmes are much more effective when women are fully involved in decision-making. The development of effective, sustainable water initiatives in rural India is vital to the country's future and the empowerment of women. Reliable access to clean water and sanitation allows Indian women to realise their potential and role in their communities and live fuller lives. Women are not a homogenised group; they are also subject to class, caste, religion, etc. Therefore, gender interventions need to reflect this too and take stock of the required ingredients to achieve this—for example, the four Es and four Cs outlined by Wani et al. (Chapter 6). The challenges and opportunities are clear and so now is the time for sustained action. We present clear action points for improving gender outcomes around water and sanitation with a focus on quality, scale and sustainability. While many of these have been touched upon in the case study chapters, we also present additional reflections that also need inclusion.

16.2 The Way Forward: Lessons from This Volume

The tragic events of December 2012 when a young student was brutally gang-raped on a moving bus in Delhi city centre shocked India and reverberated around the world.[1] The country mourned when she succumbed shortly afterwards to the horrific injuries inflicted upon her. India introspected and questioned what kind of a society has been allowed to grow up around us that such an evil event could happen? How could a young innocent girl simply coming home from the movies suffer such a tragic end? The resultant protests and soul-searching brought India to face the fact of the skewed gender norms in operation. There was a sense of urgency and a need for change—the question seemed to move from why gender to how gender. Here, we explore this 'how gender?' question. From this collection and discussion with the authors and sector practitioners, we draw out key areas for each of the critical stakeholders to address.

How should government money on social schemes be spent for gendered outcomes? Gender budget cells, established in all ministries by the Government of India, aim to influence and effect a change in policies and programmes to tackle gender imbalances and promote gender equality and development (Ministry of Finance, 2007). These cells also ensure that public resources through the ministry budget are allocated and managed with a gender-sensitive lens. This initiative, gender budgeting, views the

Government budget from a gender perspective in order to assess how it will address the different needs of women (Ministry of Women and Child Development, 2007). But only with strong monitoring and focus on evidence can this excellent initiative achieve its goal. This will help to move away from the viewpoint that centralised schemes in India, including water and sanitation, can often result in disempowering community participation by taking away the locus of control from the community to a technological fix. The WASH sector can start by examining public expenditures through a gender lens as well as undertaking beneficiary assessments around service delivery. Sufficient and robust evidence and practical application of gender responsive budgeting (GRB) is still not at hand for more widespread application in the sector in India—further application and evidence is required to show its true impact. This will require investment on the process side also and to get away from the traditional hardware approach with poor communication and consultation behind it—for example, public money spent on community toilet blocks with an explicit focus on women and girls only for such facilities to rapidly fall into disuse due to poor operation and maintenance.

There is a pressing need for stronger data that can be disaggregated to highlight the key gender components. Such data are imperative for future policy strengthening and Government must play the key role in its collection. Progress has been made in terms of some social groupings such as STs and SCs but other groupings such as gender, female-headed households and those with a disability need much more evidence to inform policy. Then, based on this evidence, policy can be made much stronger in terms of gender inclusiveness within the current programmes of Government of India. Policies and institutional reforms should go hand in hand to achieve the required results.

Institutionalising how to adopt and track the key minimum interventions is central to improved gender outcomes—for example, participation of women with quality in planning, operation and management mechanisms is vital. In addition to policy, proof of concept may be tested in terms of water and sanitation and gender programming before moving to scale. This can inform the achievement of systematic realisation of gendered outcomes, empowerment, sustainability, etc. This, and robust monitoring of the gender-sensitive components of WASH programmes [e.g. self-help groups (SHGs), water user associations] can help create opportunities for women in the sector (e.g. training of women masons). From this can come strategies for scaling-up through intra- and inter-convergence of different programmes and schemes within the government bodies and with civil society and academia, as Bastola (Chapter 7) emphasises.

SHGs have been involved in sanitation-related activities in different parts of the country as producers of sanitary napkins, latrine hardware and other products but in addition to opportunities to strengthen themselves financially, they can also offer the 'last-mile' connectivity to educate communities on hygiene and hand-washing in particular. Positive experiences abound. For instance, in Chamoli, Uttarakhand, a federation of SHGs called Alaknanda Ghaati Shilpi Federation involved in the crafts has developed a low-cost liquid hand wash. It comprises soap nuts, cow dung, apricot and orange peel. It promotes this with its members for hand-washing, and *vaids* (traditional medicine practitioners) talk about the need for hand-washing. Another type of low-cost hand-washing medium uses ash from cow dung cake mixed with soap nut powder (Solutions Exchange, 2010).

Such initiatives also lead to gender empowerment—a core result from good gender work. The key finding from this work is that programmes employing structured approaches aiming to impact equity and gender outcomes and/or sustainability must be planned and resourced for a project cycle of 2 to 5 years (Lala et al., Chapter 2). This is borne out of a broader WASH and inclusion work done by *Gram Vikas* in Odisha, which estimates that from inception to village community taking over operations of a sustainable village water supply and sanitation system is a 3- to 5-year process (UNICEF, 2013a). Yet, government interventions still typically plan and implement over a 1-year cycle (Cronin and Burgers, 2011).

Government institutions (public health centres, schools and *Anganwadi* centres) can do much more in terms of gender-friendly infrastructure. Bathing spaces along with toilets, toilets for women with special needs and disabilities, and toilets with management facilities for pregnant women and menstrual hygiene are needed. Women giving birth in rural public health centres often do so with no running water, with no soap for hand-washing, with no functioning toilet and with no proper place for placenta disposal. It is not difficult to imagine how such facilities look and smell; equally, it is not too difficult to imagine the link with the fact that half of the children who do not survive to their fifth birthday die within the first month of birth (SRS, 2011). The hardware (latrines, incinerators and bathing spaces) will only be successful with the right software (participation, community dialogue, management and maintenance, ownership, communication for behaviour change) behind it.

16.2.1 Lessons and Gaps

Gender participatory processes, such as the ones outlined by Lala et al. (Chapter 2), must be integrated into government programmes. Even timing the interventions and designing the timetable so that women can attend has to be factored in (e.g. Saxena et al., Chapter 14); still such innovative approaches need to be sustained and frequently followed up and not become just one-off events. The quality and quantity of government staff needed to achieve this must be estimated—not just for the present but also for the sector where it wants to be in 10 years from now. Capacity building is key to this: Sinha (Chapter 5) outlines the need for a review of curricula for comprehensiveness of coverage of gender issues in WASH programmes, and more comprehensive analyses are still needed that also look at filling the future increased capacity gaps, perhaps in terms of human forecasting approaches. Strategic and focused interventions around communication for sustained behaviour change and redress of skewed gender norms can support this. Proof-of-concept across states can pave the way for impact at scale.

Perhaps affirmative recruitment strategies within Government are required to achieve gender balance not just in absolute numbers but also at all levels of management. Other initiatives include career outreach, scholarships for study in the sector, career breaks and other incentives. Ultimately it is about accountability and good governance and there should be ways to achieve this by the inclusion of gender sensitivity and impact in the evaluation of government employees.

Civil society organisations working on WASH also must recognise the importance of grassroots women's leadership and how to nurture and capture these approaches, inputs and results in the gender and WASH domain. This collection is a modest start in that direction. Organisations can achieve this via first looking at their own gender policies, practices, staffing patterns, programmes and communication on WASH issues to enhance gender sensitivity. In addition, they may work on improving the organisational understanding of theory of change and up-skilling on gender-related methodologies with standardised terminology to ensure spread of results. This will require an expansion of research on gender and WASH, specifically action research, participatory research and process documentation, which would be a welcome development. Collective community action is also needed to address skewed gender norms. Women's increasing collective demands for rights and services will influence this. The key role of women in marginalised communities can be better captured and appreciated also at that level.

In general, while the water and gender discourse at least exists, the sanitation and gender discourse is less developed, and gender and hygiene is still a big gap that needs further work. 'Women have, by far, the most important influence in determining household hygiene practices and in forming habits of their children' (Jha in ADB, 2009). Limited evidence still exists in the Indian context on gender approaches to improved hygiene and the related impact (even this collection is light on hygiene despite a real effort to address this) and this must be tackled. Hence, looking forward holistic and convergent approaches that incorporate hygiene are also required.

While it is always difficult to prioritise when so many issues are important, however, an attempt has been made to do so in terms of three suggested priority actions, each with immediate, medium and high time scales (see Table 16.1).

16.3 Additional Key Actions Required

As we had outlined in the Introduction and again in the preceding sections, urgent action is needed. However, urgent action alone will not result in three critical outcomes: gender outcomes at scale, gender outcomes that have quality and finally gender outcomes that will be sustained. A key action-orientated agenda is needed that can take this dialogue forward in India to address scale, quality and sustainability. We revisit here two of the critical challenges to achieve this—climate change and on-ground programme implementation.

Table 16.1:
Potential priorities for gender in WASH interventions with timeline

Timeline	*Potential Priorities in Terms of Impact at Scale*	*How to Implement the Identified Priorities*
Immediate	• Recognition of grassroots women's leadership in WASH • Communication campaigns on Water, Sanitation and Hygiene and Gender integration • Development of guidelines for convergence across Ministries with key indicators to track progress for strengthened gender outcomes in WASH in the National Government water and sanitation programmes (NBA/NRDWP)	• Government and civil society can recognise the outstanding WASH female champions (such as the new brides refusing to go to their husbands houses until toilets were built) • Integrated Ministry/Departmental approaches to communication on WASH and gender issues • Common review and oversight taskforceresponsible also for monitoring performance and setting accountability
Mid-term	• Gender-focused proof-of-concepts with commitment to scale-up by Government • Recommendations to strengthen monitoring systems for gender-disaggregated data • Organisations reflect gender norms in their polices	• MDWS and State Govt. committing to testing process tools with gender outcomes both to show the feasibility but also to build internal capacity • Census and NSSO examine survey questions from gender perspective and ensure they will get adequately disaggregated data on WASH and gender • Civil society lead the way in terms on progressive gender policies and implementation
Longer-term	• Review of curricula for WASH and gender issues (e.g. MHM) • Government policies and accountability mechanisms and recruitment policies for gender balance • Inclusion of gender sensitivity in the evaluation of government employees	• Ministry of Women and Child Development along with Ministry of Human Resource Development (Education) would need to take responsibility here and address MHM curriculum and hardware gaps in schools (sanitary towel bins, incinerators, etc.) • The issue of gender-sensitive accountability, recruitment and performance appraisal needs to be looked at for both IAS and state civil service cadres

Source: Themes from this collection as well as discussions at the workshop 'Women-led Water Management: strategies towards water sustainability in rural India', held during 5–6 November 2012 at Gurgaon, India.

In India, research assessing climate change and its impacts on human lives have clearly indicated that a changing climate is manifesting itself in the form of unprecedented changes in almost all major ecosystems (Hallegatte et al., 2010; Mohan and Sinha, 2009) with ramifications in terms of societal impact falling heavily on the poorest. However, from the experiences outlined in this volume, we know that burden among the poorest will fall disproportionally on women.

This stark reality also presents opportunities to do gender-proof climate change interventions to ensure quality and sustainability. An important concept in this discourse is resilience. This has been defined as where adaptive capacity relates to the ability to influence and respond directly to processes of change (to shape, create or respond to change), resilience is the ability to absorb shocks or ride out changes (OXFAM, 2011). Interventions from this viewpoint present an opportunity to build capacity and empower women and children to identify and address the root causes of vulnerability and risk leading to stronger and more capable individuals, households and communities and can go beyond Disaster Risk Reduction work alone. Such an approach is by its nature multi-sectoral and tries to build on strategies in communities (UNICEF, 2013b) that result in:

- Diversified income base and livelihood strategies, including access to markets and information;
- Access to assets (financial, social, human, physical, natural) and use of flexible, quality basic social services able to adapt to shocks;
- Access to social protection programmes, including safety nets especially during difficult periods;
- Responsive and inclusive institutions/structures that address changing realities of communities and families;
- Access to information and skills that enable positive adaptive behaviours in response to shocks.

Resilience work is based on the principle of putting vulnerable households and communities first within a grounded understanding chronic vulnerability and risk. Linked to this is the need to build at higher levels of capacity and accountability around a longer-term focus that bridges humanitarian and development interventions.

The third element to address is scalability and this requires a strong government, though there is a governance deficit in the WASH landscape in India (UNICEF, FAO and SaciWATERs, 2013). The system at all levels needs a strong focus on results, especially on inequities. Currently, one of the key challenges in reaching the unreached in India is the large, specific targeting of the marginalised. Though there is a provision in budget allocations to do it, it is normally a stipulated amount, for example, 5 per cent for a certain group, as opposed to a ring-fenced budget allocation that is monitored in terms of expenditure and impact. Hence, the first critical point is to reach specific gender outcomes in the sector, with dedicated gender funds to be spent first and not last and with oversight and associated reward or punishment. To achieve the type of resilience outcomes outlined previously will require strong inter-departmental cooperation with shared programmes and one-result framework. The proliferation of schemes in India all assumes that other schemes will do this but in reality this does not happen.

Hence, in practical terms, for improving gender in WASH in India and knowing the modus operandi of the Indian political and bureaucratic systems, specific actions are needed:

1. A national task force sitting within MDWS dedicated to gender and equity results; the role to ensure capacity building and guidance to the state implementers on process.
2. Independent monitoring of results against a defined framework of indicators with a clear associated accountability framework, at all levels from village health frontline worker to Union Minister
3. Ring-fenced priority budgets that focus on quality as opposed to merely increasing coverage figures.
4. Strong political support is required but given the rise of the anti-corruption and pro-poor agenda in India this there is no reason why this could not happen if real interest for change is there.
5. Cronin and Thompson (in press) propose that to achieve points 2 and 3 requires the dove-tailing of three types of check-and-balance systems, which currently are either not fully enforced and certainly never integrated—these are linking social audits, financial audits and technical audits to ensure quality outputs at local level that reach the most needy and do so in a cost-effective manner.
6. This also would necessitate convergence not just across the various Government actors but also push Government-civil society dialogue which remains fraught at times in India but is essential

to healthy debate, improving governance and good development. Indeed, without such dialogue there is no hope to shake the deeply engrained and depressing gender norms that currently dominate.

7. Institutional reshaping and strengthening is urgently required with a much stronger institutional and policy regime for the myriad of new challenges that India's water sector is facing along with the still unresolved issues of gender and social equity. This must result in a management and enforcement regime that is evidence-driven (Cronin et al., 2014).

16.4 Final Words

Gender is a key element in the Planning Commission's vision:

> Inclusive growth should be reflected in the improvement of provision of basic amenities including water and sanitation. Particular attention is required to meet the needs of the SC/ST and OBC population.... Women and children constitute 70% of the population and deserve special attention. Ending gender-based inequities faced by girls and women must be accorded the highest priority. (Planning Commission, 2011)

We hope that this collection inspires further debate on WASH and gender in India and provides a more nuanced understanding of women in WASH issues. As mentioned in the Introduction, without addressing the fundamental underlying issues, including data, norms, institutional culture, capacity, etc., well-meaning gender inclusion efforts will not lead to logical and intended outcomes. It is critical to acknowledge women's opinion, action and steering in WASH and development and the need to start to do this at scale.

This collection shares experiences from across India that achieved success despite systemic deficiencies and weaknesses. Up-scaling of these achievements in terms of gendered outcomes in the WASH sector is possible and given the recent introspection within India around skewed gender norms the time is ripe for creating momentum at an all-India scale. However, moving to scale requires political support, strengthened participatory processes, strong monitoring backed up by gender disaggregated data and solid documentation. This all will require much stronger convergence across all stakeholders involved: Government (from policy makers to front-line workers), civil society, community organisations and society itself. Such a coalition can work together to show that these crucial results are achievable and, in doing so, the full value of WASH resources can be felt in India in terms of improved public health and empowerment, equality and education. This has yet to be achieved but can be within reach with concerted effort.

Disclaimer

The views expressed herein are those of the authors and do not necessarily reflect the views of UNICEF, the United Nations, SaciWATERs, ICIMOD or S M Sehgal Foundation.

Note

1. For more information on the 2012 Delhi gang rape, see http://en.wikipedia.org/wiki/2012_Delhi_gang_rape_case (accessed on August 2013).

References

Asian Development Bank (ADB). 2009. India's Sanitation for All. How to Make It Happen? Philippines, ADB.

Cronin, A.A. and Burgers, L. 2011. Water Safety and Security—Challenges and Opportunities in the Indian Perspective, in Confluence of Ideas and Organisations, International Conference on Water Partnerships towards Meeting the Climate Challenge, Chennai, India, January 2011, pp. 45–52.

Cronin, A.A., Prakash, A., Priya, S. and Coates, S. 2014. Water in India; situation and prospects—The results of a recent systematic sector review and consultation, *Water Policy*, 16: 425–441.

Cronin, A.A. and Thompson, N. (in press). Strengthening data and monitoring in the Indian water and sanitation sector. *Journal of Water, Sanitation and Hygiene for Development.*

Hallegatte, S. et al. 2010. 'Flood Risks, Climate Change Impacts and Adaptation Benefits in Mumbai: An Initial Assessment of Socio-Economic Consequences of Present and Climate Change Induced Flood Risks and of Possible Adaptation Options', OECD Environment Working Paper, No. 27, OECD Publishing.

Joshi, D. 2005. 'Misunderstanding gender in water: Addressing or reproducing exclusion', in A. Coles and T. Wallace (eds), *Gender, Water and Development*, Ch 8. Oxford, UK: Berg Publishers.

Kulkarni, S. 2011. Women and decentralised water governance: Issues, challenges and the way forward. *Economic and Political Weekly*, xlvi(18): 64–72.

MDWS. 2011. Strategic Plan 2011–2022 of the Ministry of Drinking Water and Sanitation Ministry of Drinking Water and Sanitation, Government of India.

———. 2012. Sanitation and Hygiene Advocacy and Communication Strategy Framework 2012–2017; Ministry of Drinking Water and Sanitation, Government of India and UNICEF, 76 pages.

———. 2013. National Rural Drinking Water Programme: Movement towards Ensuring People's Drinking Water Security in Rural India; Guidelines 2013 update; Ministry of Drinking Water and Sanitation, Government of India.

Ministry of Finance. 2007. Charter of Gender Budget Cells; Ministry of Finance, Department of Expenditure, Government of India, 8 March 2007.

Ministry of Women and Child Development. 2007. Gender Budgeting Handbook for Government of India Ministries and Departments, 78 pages.

Mohan, D. and Sinha, S. 2009. *Vulnerability Assessment of People, Livelihoods and Ecosystems in the Ganga Basin*. Delhi, WWF.

OXFAM. 2011. Introduction to Climate Change Adaptation: A Learning Companion; Oxfam Disaster Risk Reduction and Climate Change Adaptation Resources, 16 pages.

Planning Commission. 2011. Faster, Sustainable and More Inclusive Growth: An Approach to the Twelfth Five Year Plan (2012–17). Planning Commission, Government of India.

Solutions Exchange. 2010. Consolidated Reply on Role of Self Help Groups in promoting Hand Washing Experiences; Referrals issued on 20 December 2010. Available at ftp://ftp.solutionexchange.net.in/public/wes/cr/cr-se-wes-12101001.pdf (accessed on August 2013).

SRS. 2011. SRS Statistical Report 2011, Ministry of Home Affairs, Government of India.

UN. 2012. Stigma and the Realization of the Human Rights to Water and Sanitation, Report of the Special Rapporteur (Catarina de Albuquerque) on the human right to safe drinking water and sanitation, United Nations General Assembly, Human Rights Council A/HRC/21/42 July 2012.

UNICEF. 2013a. Lessons in Inclusive Programming, Reducing Disparities from Interventions in the WASH Sector in India, UNICEF Policy, Planning and Evaluation (PPE) Programme.

———. 2013b. UNICEF and Resilience—Briefing Note for Senior Staff, 2 pages.

UNICEF/FAO/SaciWATERs. 2013. Water in India: Situation and Prospects, UNICEF, New Delhi, India. Available online at http://www.unicef.org/india/media_8098.htm (accessed on 8 April 2013).

Zwarteveen, Margreet, Ahmed, Sara and Gautam, and Suman Rimal. 2012. *Diverting the Flow: Gender Equity and Water in South Asia*. New Delhi, Zubaan Publishers.

About the Editors and Contributors

Editors

Aidan A. Cronin trained as a civil and environmental engineer and holds a PhD in Water Resources from Queens University, Belfast. He has worked in consultancy and then as a Senior Research Fellow at the Robens Centre for Public and Environmental Health, University of Surrey, UK, where he spent five years researching the impact of anthropogenic activities on water quality in the EU and developing country settings. He then worked as a water and sanitation advisor at the United Nation High Commissioner for Refugees in their Public Health Section in Geneva, Switzerland, before joining UNICEF India in 2008. He managed the UNICEF water and sanitation programme in Odisha, India, up to September 2010 when he joined the New Delhi office as the water advisor. His research interests are in understanding the impact and contribution (health, nutrition, economic and social) of WASH provision and the processes needed to achieve these.

Pradeep K. Mehta is Group Leader, Rural Research Centre, S M Sehgal Foundation, Gurgaon, India. He holds a PhD in Economics from Mysore University through the Institute for Social and Economic Change (ISEC), Bangalore; an MPhil in Planning and Development from Indian Institute of Technology (IIT), Bombay; and M.A. and B.A. degrees in Economics (honours) from Punjab University, Chandigarh. A development specialist, he has over eight years of experience in teaching and research. His areas of expertise are rural development, agriculture, climate change gender, water and impact evaluation.

Anjal Prakash is the Executive Director at SaciWATERs, South Asia Consortium for Interdisciplinary Water Resources Studies based at Hyderabad in Southern India. He is also the Project Director of 'Water Security in Peri-Urban South Asia,' a project funded by IDRC. He has worked extensively on the issues of groundwater management, gender, natural resource management, and water supply and sanitation. Having an advanced degree from Tata Institute of Social Sciences (TISS), Mumbai, India, and PhD in Social and Environmental Sciences from Wageningen University, the Netherlands, Dr Prakash has been working in the area of policy research, advocacy, capacity building, knowledge development, networking and implementation of large-scale environmental development projects. Before joining SaciWATERs, Dr Prakash worked with the policy team of WaterAid India, New Delhi, where he handled research and implementation of projects related to Integrated Water Resources Management (IWRM). Dr Prakash is the author of *The Dark Zone: Groundwater Irrigation, Politics and Social Power in North Gujarat*, published by Orient Longman. His recent edited books are *Interlacing Water and Health: Case Studies from South Asia* (2012) by SAGE Publications and *Water Resources Policies in South Asia* (2013) by Routledge. He is presently co-editing books on case studies of IWRM and Peri-Urban Water Security Issues to be published by Routledge and Oxford University Press, respectively.

Contributors

Satyabrata Acharya has done Master's in Agriculture in 1989. He joined Professional Assistance for Development Action (PRADAN) in 1990. For over a decade, he had worked for grassroots mobilisation of rural families around the issues of financial inclusion, food security and livelihoods. He is currently based at Ranchi, integrating the operations of PRADAN in the state of Jharkhand, spread over 14 districts covering 1.25 lakh families.

K.H. Anantha is working as a Scientist in Resilient Dryland Systems Programme of International Crops Research Institute for the Semi-Arid Tropics (ICRISAT) since 2009. His areas of interest are Economics: Land and Water Resources, Groundwater Management, Tank Management, Agriculture Development, Climate Change, Development Economics and Rural Development. He holds a PhD in Economics from University of Mysore, Mysore, through Centre for Ecological Economics and Natural Resources (CEENR), Institute for Social and Economic Change (ISEC), Bangalore, India. He is a holder of Young Professional Research Fellowship

(YPRF) from International Water Management Institute (IWMI) as well as Indian Council of Social Science Research for his PhD.

Ch. Ram Babu is the Executive Director of an NGO, *Grama Vikas Samstha*, working in Pungannur and Madanapally for the last two decades. He spent his early professional life with Gandhi Peace Foundation and continues to be in its board. He has been involved with groundwater management by communities over the last decade and is currently offering field-level exposure training to World Bank–supported projects from all over the country.

Aditya Bastola holds a PhD in women's studies and has varied area of interests in gender, water and climate change research and development. He has conducted independent researches in South Asia. He was also associated with the S M Sehgal Foundation, Gurgaon, as a social scientist.

Malika Basu holds a PhD in Development Studies from the Institute of Social Studies, The Hague. A social development specialist and gender analyst, she has over 18 years of experience in research, evaluation and policy analysis covering rural, social and gender issues. Starting with MARG, a Delhi-based NGO, she later worked in project support and advisory roles for national and international organisations on issues of participation and inclusion of marginalised communities, women's empowerment, self-help groups, livelihoods, development-induced displacement and related problems. In 2008, she joined the Solution Exchange initiative of the UN Country Team in India to head the Gender Community of Practice, undertaking knowledge management to provide knowledge-based services to development organisations and practitioners.

Somnath Basu is working in UNICEF for more than 12 years and served three different state offices in West Bengal, Assam and currently Jharkhand. Dr Basu has worked intensively with communities and government systems for building institutional mechanisms to deliver water and sanitation programmes. Some of the critical milestones of Dr Basu's work have been to promote water and sanitation within excluded communities. Dr Basu did his PhD in Urban Geography from Jawaharlal Nehru University, New Delhi.

Tapas Chakma, MBBS, MAE, has been working as a senior scientist at the Regional Medical Research Centre for Tribals, Jabalpur, for the last 23 years. He is an expert in dealing with different aspects of fluorosis, that is, health-related issues, nutrition and water-related aspects. He has provided his expertise not only in India but also abroad. He is recognised as a fluorosis expert by the Ministry of Health and the Ministry of Drinking Water Supply.

Chanda Gurung Goodrich is the Senior Gender Specialist at the International Center for Integrated Mountain Development (ICIMOD), Kathmandu, Nepal. She specialises in participatory research and development, and gender, particularly in the area of natural resources management. She has worked extensively on issues concerning social and gender equity rights, particularly in the spheres of agriculture, natural resources management and sustainable livelihoods in India and Nepal since 1996. She has published several papers and chapters in edited volumes on gender and agriculture/natural resource management.

Megha Jain has been working in the water and sanitation sector in rural and urban areas for the last seven years. She has been undertaking research and documentation projects with UNICEF, Water and Sanitation Programme (The World Bank), Public Health Foundation of India, CRISIL, Feedback Foundation, Infrastructure Professionals Enterprise, Centre for Urban and Regional Excellence, National Institute of Public Finance and Policy, etc. Some of her recent projects include study on Best Practice and Lessons Learnt from CLTS in Madhya Pradesh; Monitoring, Evaluation and Documentation for Rural Sanitation Team at WSP for ongoing work on Total Sanitation Campaign for creating open defecation–free communities; Rapid Baseline Assessments for Cities on Urban Governance and Service Delivery; Appraisal of Reforms under JNNURM amongst many others. Sectorally, she has worked in areas of water and sanitation, governance and reforms, and urban development. Her skill areas include research, documentation, assessments, project implementation and appraisal, business development and networking.

Jyotsna is a postgraduate in Development Studies from Tata Institute of Social Sciences. She has worked with Centre for Equity Studies on the issue of Bonded Labour and has experience of engaging in movements, campaigns and seminars, and has conducted several workshops with adolescent girls in several villages of Osmanabad district, Maharashtra. She is deeply interested in conducting feminist research and strengthening participatory research methodology especially in the areas of gender, caste, performance and intimacy. At present, she is working as a gender consultant with the Ministry of Drinking Water and Sanitation, New Delhi.

Yusuf Kabir has been working in sanitation and water sector for the last 13 years. Yusuf has been with UNICEF since 2007. Prior to that he has worked with organisations like DFID, national-level NGOs, social and marketing research consultancy firms like GFK-MODE, ORG India Pvt. Ltd, Ramky Infrastructure, SREI Capital Markets and SPAN Consultancy on issues related with environment and livelihood development training and capacity building in the social sector, etc. Yusuf is a commonwealth scholar and a trained policy writer from Central European University, Budapest, Hungary, where he had undergone a summer course on 'Evidence Based Policy Formulation'. Yusuf holds postgraduate degrees in Sustainable Development from Staffordshire University, UK, and Public System Management with a specialisation in Environment Management from Indian Institute of Social Welfare and Business Management, Calcutta. Along with TISS, Mumbai, Yusuf has developed India's premier postgraduate course on WASH.

Eshwer Kale is working with Watershed Organisation Trust (WOTR), Pune, in Climate Change Adaptation Project and pursuing his doctoral study on groundwater governance issues at Tata Institute of Social Sciences (TISS), Mumbai. He is also involved with various advocacy and grassroots-level groups working on various water issues. He completed his MPhil from TISS and Master's in Social Work from Karve Institute of Social Services, Pune. He was associated with SOPPECOM (Society for Promoting Participatory Eco System Management, Pune) for three years in various research projects on watershed issues.

M. Dinesh Kumar did his BTech in Civil Engineering in 1988, ME in Civil (Water Resources Management) in 1991 and PhD in Water Management in 2006. He has nearly 22 years of professional experience in the field of water resources, in undertaking research, action research, consulting and training in technical, economic, institutional and policy issues related to water management with several prestigious national and international organisations, including Institute of Rural Management, Anand, and International Water Management Institute, Colombo. Since August 2008, he is Executive Director of the Hyderabad-based Institute for Resource Analysis and Policy, which is a non-profit research organisation engaged in interdisciplinary and multidisciplinary research in the field of natural resources, which he created. He has nearly 135 research publications to his credit, including three books, several book chapters, many articles in peer-reviewed journals and research monographs. He is the author of three books. His most recent book is titled *Managing Water in River Basins: Hydrology, Economics and Institutions*, published by Oxford University Press, New Delhi.

Nisheeth Kumar is the founder member and Chief, Operations, Knowledge Links, a consulting outfit specialising in capacity development for community and people led change with its domain expertise in the water and sanitation sector. His special interest and skill lie in designing and implementing capacity development interventions on scale. He works on issues related to water and sanitation, disaster risk reduction, sustainable households and health. The functional domains that he works in include training, research, monitoring and evaluation, and knowledge management.

Sunetra Lala holds a postgraduate degree in Environmental Management and Environmental Law and a graduate degree in Biological Sciences (Zoology Honours). In the past, Sunetra has worked with the UNDP, Afghanistan with the Communications Unit. Prior to this, she led a team on Corporate Sustainability Management (CSM) with the Confederation of Indian Industries. Sunetra has also been an Editor for the Penguin Publishing Group India, and has worked with the Education Department of Sikkim, Centre for Environment Education, and Centre for Science and Environment.

K.A.S. Mani has over three decades of experience working in India, Afghanistan, Cambodia, Vietnam, Sultanate of Oman and Mozambique on developmental issues largely centred around water, as well as on management of humanitarian response to drinking water during emergencies. He has implemented groundwater component of World Bank-funded Hydrology Project in several states, and has been involved with the design of groundwater database and automated water level recordings. He has participated in Guided Research to Government departments, universities and NGOs on issues related to water resource management.

Gregor von Medeazza has around 10 years of work experience in the WASH sector and has been heading the WASH programme in UNICEF State Office for Madhya Pradesh, India, for the past three years, fostering close and innovative partnerships with Government, leading CSOs and research institutions, the private sector as well as international partners. Before his current position, Dr von Medeazza was UNICEF WASH Specialist in Mali, where he led the country's efforts in the area of Handwashing promotion and contributed to the CLTS, WASH-in-Schools and HWTS programmes. Dr von Medeazza started his career with UNICEF in its New York headquarters and supported the WASH Cluster in Haiti, shortly after the 2010 devastating earthquake. Dr von Medeazza's background includes a PhD in Ecological Economics and Environmental Management, during which he worked on the implementation of water provision systems under freshwater scarcity and conflict situations in the Maghreb region and India, and with UNDP in the West Bank and the Gaza Strip. He also holds a Master's in Civil and Environmental Engineering from Imperial College, London, for which he conducted a feasibility study on solar-driven water pumps in rural Bolivia. Dr von Medeazza completed executive trainings at Harvard University and SAIS (Johns Hopkins University), and is currently a part of INSEAD's Leadership Development Programme.

Amit Mehrotra has been working in the development sector for over two decades now. After obtaining his MTech from IIT-Roorkee, he started his career with Action for Food Production as Water Resource Specialist in the year 1993. Since then, he has worked in various national organisations, NGOs and bilateral programmes. Currently, he is working as WASH Specialist with UNICEF. He has worked in different sectors viz., Drinking Water and Sanitation, Watershed Management, Natural Resource

Management, Emergency, RCH and Livelihoods. He is providing leadership to UNICEF supported WASH programme in the largest state of the country, Uttar Pradesh.

Pradeep Meshram, MA, MPhil, is working as a Technical Assistant (Research) at Regional Medical Research Centre for Tribals, Jabalpur, for the last 17 years after successfully working at National Nutrition Monitoring Bureau for five years.

Shailesh Mujumdar is a development professional with 23 years of experience in 'for profit,' 'government' and 'not for profit' sectors. He did postgraduate diploma in Management of 'non-government organisations' from Entrepreneurship Development Institute of India (Ahmedabad). During the course, he won Silver medal and was also awarded with Sir Ratan Tata fellowship. His competencies are in Monitoring and evaluations, Project management, Capacity development, Research, Advocacy and Campaigns and Community-based processes.

Sudhir Prasad is currently the Additional Chief Secretary of the Drinking Water and Sanitation Department, Government of Jharkhand. An IAS officer of 1981 batch, Mr Prasad did his BTech in Mechanical Engineering from IIT, Kanpur, a certificate course conducted by Harvard University and two certificate courses conducted by IRC, Netherlands. Mr Prasad has served as Secretary, leading various departments like Science and Technology, Energy, Environment & Forest and Planning, etc. of the Government of Jharkhand. During the span of 32 years of service, Mr Prasad as administrative head of districts and departments at the state level has taken active interest in understanding issues of the tribal population and design programme outreach to address their specific social and cultural needs.

Vallaperla Paul Raja Rao, executive director of an NGO, BIRDS, has over the last 25 years worked with the poorest of the poor, educating them with knowledge, skills and capacity to help them support themselves. He has implemented several projects in the area of community-managed borewells, community management of groundwater and climate change. He is associated with several NGO networks and is a member of several committees of Government of India, and has won several awards.

Madhukar Reddy is a Doctorate in Botany, specialised in ethnobotanical studies. For more than two decades, Dr Reddy has been actively involved with NGOs for the development of resource-poor farmers. He established a voluntary organisation, SAFE (Society for Sustainable Agriculture and Forest Ecology), with a mission of enabling the people to adopt sustainable, environment-friendly, equitable and cost-effective livelihood systems. He is actively involved in training the farmer on ecological issues focusing on afforestation, regeneration and sustainable utilisation of forest resources. He has also served as a principal investigator for Government of India projects—waste land development and medicinal plant cultivation. Presently, he is actively involved in projects such as Farmer Ground Water Management, Climate Change—Farmer Adaptations, and Organic Agriculture. Also, he is involved in training farmers in cultivation, post-harvest and processing techniques of minor forest produce. As a consultant, he was a Mission Member to develop Master Plan for Degraded River Basin in The Royal Kingdom of Bhutan.

Ajit K. Saxena (MTech, PGD in Environment Management and Sustainable Development) is a WASH rural and urban Infrastructure and Renewable Energy Expert. He brings 24 years of experience at varied levels. He has worked with a government agency and a national-level non-government organisation and associated with UN agencies like UNDP and UNICEF in India and UNESCAP in Bangkok. He was consultant to Asian Development Bank (researcher on sanitation), sanitation expert to facilitate trainings (Water Sanitation Programme, the World Bank-STEM), team leader of TA on Septage Management Plan for urban WASH for ADB, team leader, WASH scoping study for DFID for MP, Advocacy/Validation reports and planning for WaterAid. He is a founder member of 'EEDS', a state-level NGO on WASH, which initiated 'NGO secretariat' to provide valuable support to 110 grassroots organisations and development agencies.

Niti Saxena is working as a Senior Scientist at the Rural Research Center of S M Sehgal Foundation. She has development experience in Asia while working across sectors for non-governmental organizations, the United Nations and international donor agencies. She holds a master's in Human Development from Lady Irwin College, Delhi University. She has over nine years of experience in social science research. Her areas of expertise include qualitative research and participatory development.

Janardan Prasad Shukla is the founder member and chief, Policy and Programmes, Knowledge Links, a consulting outfit specialising in capacity development for community and people led change with its domain expertise in the water and sanitation sector. He is a leading champion of CLTS in India and has been instrumental in training a large number of CLTS master trainers and facilitators in the states of Andhra Pradesh, Assam, Chhattisgarh, Himachal Pradesh, Karnataka, Madhya Pradesh, Maharashtra, Meghalaya and Uttarakhand. He has been engaged in facilitating the establishment of a locally registered organisation of natural leaders from open defecation (ODF) villages in Guna block of Guna district in Madhya Pradesh.

Ajay Singh has been working in the development sector (both in rural as well as urban areas) for the last 14 years in the states of Rajasthan, Bihar and Uttar Pradesh. Recently, he completed his assignment with UNICEF office for Uttar Pradesh as Consultant—WASH in Schools. A graduate from Indian Institute of Mass Communication, he has also worked as a video journalist and provided content to various TV programmes and was part of quite a few documentaries for Doordarshan and other channels.

Sanjay Singh is a Bachelor of Civil Engineering from Karnataka University and did his postgraduate diploma in Environmental Education from Kurukshetra University. He has experience of about 15 years in water and sanitation sector. He has worked in community-based Rural Water and Sanitation project as Portfolio Manager, which was implemented in the Bundelkhand region of Uttar Pradesh and is generally known as Swajal Project. Later on, he worked as an Engineer (Drinking Water) with Department of Rural Development, Government of Uttar Pradesh and as a Specialist (Water Conservation and Community Management) with Public Health Engineering Department, Government of Madhya Pradesh. Since the last six years, he is associated with UNICEF Bhopal as State Consultant

(Water Sanitation and Hygiene). He has been instrumental in UNICEF's various interventions on fluoride mitigation during this period, which includes 'Wise water Management', 'Integrated Approach to Fluoride Mitigation' and 'Fluoride Mitigation through Nutrition Supplementation', which were implemented in fluoride-affected districts of Dhar, Jahbua and Seoni of Madhya Pradesh. He also led the zonal team and coordinated at state level for multi-district assessment of water safety in Madhya Pradesh where water quality of about 7,000 sources across the state was analysed for fluoride.

Swati Sinha is working as Freelance Social Development Consultant, based in Kathmandu, Nepal. Currently, she is pursuing MPhil from CESS (Centre for Economic and Social Studies) in Sociology. She has obtained Master's in Rural Development (PGDRD) from Xavier Institute of Social Service (XISS), Ranchi, in the year 1999. Since then, she has been working with NGOs on issues related to governance, decentralisation, community mobilisation, gender, water and democracy, access to water and sanitation services, solid waste management, participatory planning and integrated water resource management. She was associated with UNNATI, Ahmedabad, and PRIA, New Delhi, for eight years till 2008. Since then she has worked as freelance social development consultant with MARI (Modern Architects for Rural India), CWS, YFA, SOPPECOM and IPE Global, Delhi, on short-term assignments related to training manual development, resource kit and policy research issues.

T.K. Sreedevi, IAS, is an Andhra Cadre Officer and she was on deputation at ICRISAT working as Senior Scientist (Watershed Development) in Global Theme on Agroecosystems at ICRISAT. Ms. Sreedevi's core interest is sustainable rural development in developing countries. She is trained in natural resource management, participatory research and development and has extensive working experience with the communities as Project Director, Drought Prone Areas Programme and Additional Project Coordinator, Andhra Pradesh Rural Livelihoods Programme (APRLP). Her work mainly focuses on issues of community participation, equity, institution building, capacity development, gender empowerment for holistic development of the community in the watershed framework. She was a member in the team, which has won number of awards including Doreen

Masher Award (2006), Resource Mobilisation Award (2006), Award for participatory trials for enhancing water use efficiency in the farmers' fields from Ministry of Water Resources (2009). She has published 50 publications in national and international journals. She is back to her parental department and working as Commissioner (R&R), I&CAD Department, Government of Andhra Pradesh, Hyderabad.

Ajit Tiwari has been working in the rural development department of the state government of Madhya Pradesh for the last 20 years and has gathered experience of implementation of the main rural development schemes at the block level. He worked with a World Bank funded Livelihood project in the state between the years 2000 and 2005. He has been associated with the Total Sanitation Campaign and mobilised the *panchayats* and the community to stop the practice of open defecation and enabled to render over 60 villages open defecation free, through a partnership with UNICEF. He has also been working with a UN Women programme to strengthen the capacity of elected *Panchayat* women *Sarpanch* and member for their effective involvement in local self-government.

Niranjan Vedantam did his B.E. in Civil Engineering and MTech in Environment Management. He has nearly seven years of professional experience, including development of environmental management systems; carrying out environmental engineering research; and undertaking environmental impact assessment and research on water management issues. Since January 2010, he is working as a Research Officer with Institute for Resource Analysis and Policy (IRAP), Hyderabad.

Suhas P. Wani is Assistant Research Programme Director and Principal Scientist (Watersheds), in Resilient Dryland Systems at International Crops Research Institute for the Semi-Arid Tropics (ICRISAT), Patancheru, India. His area of specialisation is integrated watershed management, wasteland development, biodiesel plantation, integrated nutrient management and carbon sequestration for the conservation of natural resources and their sustainable use for improving livelihoods in the semiarid tropics. Dr Wani is a University Gold Medallist and has served as an expert to make a presentation and address the members of the Parliament Forum on Water Conservation and Management; as a member of Working

Group on Minor Irrigation and Watershed Management for the 12th Five-Year Plan; as a member of Expert Committee for technical evaluation for National Initiative on Climate Resilient Agriculture (NICRA), launched by ICAR; member of the Programme Advisory Committee for Natural Resource Management and Climate Change in MS Swaminathan Research Foundation, Chennai, India and Honorary Trustee of S M Sehgal Foundation (SMSF) and organising committee member. He has served as Sustainable Agriculture Advisory Board (SAAB) Member of Unilever. He has received National Groundwater Augmentation Award at the national level from Ministry of Water Resources; Doreen Mashler Award; Best Team Award, Best ICAR Award and Outstanding scientific article award. He has organised number of national and international meetings and co-coordinating number of multidisciplinary, multi-country projects in the area of sustainable natural resource management and for improving livelihoods and building resilience of the farmers. He has published 440 research papers, books and conference papers in international and national scientific journals and has supervised six PhD students.

Dipak Zade is an anthropologist by training and is associated with the Knowledge Management Unit of Watershed Organisation Trust (WOTR), Pune. As a researcher, he is engaged in work related to climate change adaptation in rural semi-arid regions. His area of interest includes adaptive sustainable agriculture. He was earlier involved in a project related to watershed development and child nutrition and is the co-author of a recently published book *Watershed Development and Health*. Prior to WOTR, he had worked in International Institute for Population Science (IIPS), Mumbai, and KEM Hospital Research Centre, Pune.

Index